Safety Management in a Competitive Business Environment

Ergonomics Design and Management: Theory and Applications

Series Editor
Waldemar Karwowski
Industrial Engineering and Management Systems
University of Central Florida (UCF) – Orlando, Florida

Safety Management in a Competitive Business Environment

JURAJ SINAY

CRC Press
Taylor & Francis Group
Boca Raton   London   New York

CRC Press is an imprint of the
Taylor & Francis Group, an **informa** business

CRC Press
Taylor & Francis Group
6000 Broken Sound Parkway NW, Suite 300
Boca Raton, FL 33487-2742

First issued in paperback 2017

Version Date: 20140115

ISBN 13: 978-1-4822-0385-1 (hbk)
ISBN 13: 978-1-138-07534-4 (pbk)

Library of Congress Cataloging-in-Publication Data

Sinay, J. (Juraj)
    Safety management in a competitive business environment / author, Juraj Sinay.
        pages cm
    Summary: "The book provides information necessary for systematic management of traditional as well as emerging risks in the man-machine-environment system, especially in industrial technologies. It is intended for managerial staff of various levels of management as well as product distribution workers, and for education of professionals in the fields of occupational safety and health (OSH), safety of technological systems and partially also civil security"-- Provided by publisher.
    Includes bibliographical references and index.
    ISBN 978-1-4822-0385-1 (hardback)
    1. Industrial safety. I. Title.

T55.S494 2014
658.3'82--dc23   2013049462

Visit the Taylor & Francis Web site at
http://www.taylorandfrancis.com

and the CRC Press Web site at
http://www.crcpress.com

Contents

Preface

The safety of the people shall be the highest law.

—Cicero

A safe life is not something that simply happens on its own. There is always something that needs to be done so that people feel that their environment, including their workplace, is safe—that they are not threatened by an accident, injury, or that the activities they carry out do not threaten or badly influence their health. To realize a safe life in a safe workplace and private environment means acquiring the principles of a safety culture and including them in all of one's actions. Occupational health and safety is embedded in the basic laws of economically developed countries.

If we take into consideration the fact that absolute, 100 percent safety does not exist, then we can state that zero-risk does not exist either. This philosophical approach toward life conditions the relationship between man and environment. The intensity of this relationship, as a part of safety culture, is directly dependent on having a society that is economically, ethically, and morally developed.

The economic determination of the value of human life is possible only from a very subjective point of view. This value is determined especially by attitudes of insurance companies. Most experts and scientists dealing with occupational health and safety, safety of machines and mechanical systems, safety of technological units, and civil safety, are of the opinion that an objective evaluation of human life can only be approximate. I agree with the opinion that everybody's life has the same value, and that it is important for any society to create conditions that protect human life during any activities within the environment. It is only natural that health protection is a more parametric system. One must be an active element of this system and act to include prevention of negative conditions in one's culture of life. This means that one must be convinced to live so that any risks confronted during life will be minimal or minimized.

The relationship toward a safe life can be inherited, imprinted through upbringing and education, but it will surely not be created by itself. Therefore, it is important to lay out conditions to mediate consciousness for safe actions to children from the first, which begins in kindergarten. It is only right to include classes on practical environmental safety habits for this generation's youth in elementary schools and high schools. Laying out these conditions during the first three years of college education is an effective procedure. Scholarly discussions nowadays aim at assigning specialized lectures and seminars dealing with safety and risks as part of our environment in engineering, science, and socioscientific programs. It is effective to educate top safety and risk management experts in specialized study programs at engineering colleges, while a multidisciplinary approach to the issue of safety in practical life is used therein.

The history of establishing the study program Safety of Technical Systems at Faculty of Mechanical Engineering, Technical University in Košice, is related to this trend as well. However, activity indicative of 'coincidence' preceded this trend, but since I do not believe in coincidences, I have the honour of stating, "It probably was meant to be so!"

Back in 1977, it was Professor Ing. Norbert Szuttor, DrSc, not even aware of the need to solve machine risks, who accepted me as an intern at the former Department of Transportation Machines and Facilities after my studies in the field of transportation machines and facilities. He suggested a topic related to solving safety issues when operating lifting machines and tower cranes, which are prone to safety issues due to the possibility to lose balance because of their working purpose. A load limiter had to be mounted on the crane to prevent such situations. The purpose of this device was to 'secure' the crane's stable condition, that is, to prevent its fall during operation, without intervention of the operator. However, there was a false assumption at the time on the part of every producer in the world that the lifting process was static in character, which was the basis of the principle of using load limiters. Any operational movement of a tower crane, distinct from the dynamic feature of lifting the load, was deemed a dangerous operational movement. At that time, when the notion of *risk* as a basic attribute of safety was not known in professional or in public circles, these movements presented the highest risk when operating this device. Construction of an electronic load limiter was the result of my dissertation thesis, "Dynamic Force and the Influence of Weight on Safeguarding Devices of Lifting Machines", which I successfully defended in 1978. A distinctive feature of the device was the ability to identify the real operational conditions of a tower crane and to reliably protect it from loss of balance during operation, which meant minimizing the risk of fall, and hence the threat to its operators or third parties. It proved to be unique and was patented in three countries, and was highly regarded by experts in Western Europe, namely, in West Germany, France, and the Netherlands, as well as in the German Democratic Republic.

While revising these activities in the field of technical device safety, I found that, even though not fully aware of it at that time, the beginnings of my specialized relationship to workplace safety and methods of complex risk management of technical devices began there.

In 1980, the partnership agreement between the cities of Wuppertal and Košice, which resulted in a partnership between the University of Berg, Germany, and the then named Technical College in Košice in 1982, was yet another milestone in the development of science about safety, as well as the field of study called *safety technology*. At that time, the University of Berg in Wuppertal was the only higher educational institution to have an independent Faculty of Safety Technology focused on issues of occupational safety, including safety of technical systems.

One of the invitees of the 1987 scientific conference for departments of both universities, held at Technical College in Košice (TC), was Dr Ing. Helmut Strnad from Wuppertal, a professor in the field of safe construction of machines. While looking for a partner at that time, it was agreed that the most suitable workplace that was scientifically focused on this field was the Department of Transportation Machines and Facilities of the Faculty of Mechanical Engineering. My knowledge of German seemed to come in handy during discussions with Professor Strnad at our department when we came to realize that we were actually carrying out similar scientific and professional activities. Even though we knew much about each other, the beginnings of my active cooperation were rather symbolic, since we progressed in our common activities very carefully.

Thanks to the support of my friends from Wuppertal, and especially from Ernst-Adreas Ziegler, the director of the Press and Information Office in Wuppertal, and to my prior scientific and professional activities, in 1988 the Friedrich Ebert Foundation granted me a two-year scholarship with the aim of writing an inaugural dissertation thesis at the Faculty of Safety Engineering of the University of Berg in Wuppertal. The person who accepted me into the faculty and built conditions for my research in the field that I chose as a part of my future scientific activities was the founder and first dean of this faculty, Dr. Ing. Peter Compes. I had an interesting conversation with him while I was forming the title of the project on which I would eventually work. I tried to look for new ways to construct lifting machines so they would be loaded with minimal risk during their operation. However, Professor Compes and Professor Strnad had different ideas. They had already found the most important thing for the workplace at that time—*occupational risk systematization*—something that was unknown then.

After a few days, I associated myself with the content even though I was aware that it would not be easy to work on this topic. Risks and their management systems, with the aim of risk minimization and therefore occupational health and safety, became an integral part of my everyday life. Although my family was 1486 kilometres away from Wuppertal, I was surrounded by top experts in the field of safety technology, which in general formed a unique and creative environment for my work.

After three years of intensive work, I was honoured to successfully defend my inaugural dissertation in 1990 at the Faculty of Safety Technology of the University of Berg in Wuppertal under the title of *Beitrag zur Qualifizierung und Quantifizierung von Risiko-Faktoren in der Fördertechnik dargestellt am Beispiel von Hebezeugen* (A Report on Qualification and Quantification of Risk Factors on Transportation Machines—Lifting Machines as an Example), which was published as a monograph by German VDI Düsseldorf publishing company at the end of 1990.

We began the transportation machines study program focused on the safety of technical devices shortly after my return to the Department of Transportation and Handling Technologies at the Faculty of Mechanical Engineering at the newly renamed Technical University in Košice (TUKE). The first students, admitted in 1992, graduated five years later as fresh graduates of this specialized field of study.

The independent Department of Safety and Quality of the Faculty of Mechanical Engineering at Technical University in Košice/SR was established in 2002. As in my case, risks became part of the professional life of my colleagues from the original Department of Transportation and Handling Technologies, who became members of this department after I convinced them that safety was the highest priority of any construction and manufacturing activities. Even though I am not going to mention their names, I would still like to thank them for becoming members of the scientific team that I had the honour to lead and shape.

The Department of Safety and Quality is currently a well-established scientific and educational facility with a developed scientific portfolio not only in the Slovak

Republic, but also within the European research area. Results of its work with practices in the field of applied research, as well as in the scientific lifelong learning program, justify the fact that we are focused, above all, on the needs of those who manage occupational safety and health. Mutual cooperation with other academic institutions and those from the field set up conditions for a successful process. We make the most of our cooperation in this field, especially with Germany, the United States, the Czech Republic, and others.

Several requested publications were created in this prosperous and creative environment. We have been invited to notable scientific and specialized events that have included lectures that always reveal new findings as well as specific information or drafts of specific methods and methodologies for effective risk management as a part of safety culture, which has to be a part of any modern and economically successful society.

This monograph contains a summary of my lecturing and advisory activities over twenty-five years of research, either as an independent author or as a coauthor with my colleagues, but mostly with my successful doctorate students. The monograph aims to provide the reader with an integrated and systematic point of view on most of the areas in the field of occupational health and safety management, safety of machines and machinery, certain complex technologies, and partially in the field of civil safety as a part of safety culture in the sense of national culture—a field that is now becoming very topical. It naturally follows that all the published research is done in qualifying theses, including the doctoral dissertation thesis titled "Identification, Management and Prognosis of Technical Risks of Transportation Machines and Systems", as well the first monograph in the field of risk theory, published in 1997 by OTA Košice and titled, "Risks of Technical Devices: Risk Management".

I would like to thank everyone for their support in my attempts to establish the field of safety in a complex sense as a study field of science. My thanks go to all of my coworkers, colleagues from 'my' Department of Safety and Quality of the Faculty of Mechanical Engineering at Technical University in Košice, who found a scientific school at our workplace recognized both at home and abroad. I appreciate the creative cooperation and fellowship in specialized areas with all my friends from the partner faculties and departments of foreign colleges. My supporters outside the Slovak borders definitely played a huge role in my professional life. To all of you, thank you very much!

I probably would not have been able to sacrifice a significant part of my life to scientific and specialized work if it had not been for the support of my family, my wife and daughter. Therefore, it is a great honour for me to dedicate this acknowledgment at the end of the preface to this monograph exactly to you!

Juraj Sinay
Košice

Introduction

Safe machines, environment, and the ability to perceive safety as a part of all activities people do are basic attributes of life today. If we follow the statement that 100 percent safety does not exist, and therefore risks cannot be at zero value, then it is important to prepare people for the fact that they will be confronted with dangers or hazards anytime and everywhere.

The criterion for assessing safety is a quantitative definition of risks by approximation and is one of the most important parts of their management process. Managing risk in a process involves managing all activities that are part of the process so that minimum negative effects are experienced.

Safety culture assumes that systematic measures and activities with the aim of creating a safe environment are executed within all the stages of safety management. Experts understand the notion of *safety* as *integrated safety*, which includes occupational health and safety within the man–machine–environment system, safety of technical systems, and civil safety in a broader sense.

People have started to realize that the biggest value they have is their own life and its quality, and that is where understanding the *culture of safety* principle as a part of the culture of life, and especially its use in professional and private life, originates.

Developing new technologies and materials, an aging labour force, and globalization of labour markets create conditions for new risks. Understanding this fact has to be a part of an environment where safety culture has its irreplaceable role, since it is objectively a part of people's activities. New risks do not require a specific attitude to minimize their effect. They only broaden portfolios of experts as well as the portfolios of specialists for risk management, since information and communication technologies are constantly becoming more relevant.

It is therefore held, in this sense, that man is the 'weakest link' in the man–machine–environment system. Sometimes, within the analysis of accidents or failures, it might look like the technology failed, when a detailed examination of the occurrence mechanism shows that it is the *man* who is usually the basis of the issue. Only dangers like natural disasters, such as earthquakes or weather conditions, are exceptions, although current science states that it is the people who are partially responsible even for these climate changes.

Understanding all of the relationships within risk management as a part of a culture of safety requires the active application of results from scientific and research projects, as well as their implementation in educational processes on all levels. Academic experts at various directing levels from any field of study (but mainly engineering, and the natural and social sciences) should acquire, master, and include the issues of safe life into their programs.

Occupational safety, safety of technical systems (safety), as well as civil safety (security) are, with regard to their position in society, a part of European legislation. These legislative standards presuppose their transformation into national legislative directives of the member states of the European Union. Observing these laws is binding in order to effectively execute activities within management procedures. Their effective application in practical activities requires complete knowledge of them, accepting their philosophy, and being associated with them. In this field, memorized procedures are not proof of the fact that safety culture is a part of working habits.

Globalization of labour markets and internationalization of manufacturing technologies assume a unity of binding regulations in the field of safety. Accidents, injuries, and failures have no geographical or national boundaries. Coordination of the legislative regulations is therefore mainly a part of employees' mobility strategy rather than employers' mobility strategy. Recognition of qualifications and certification of capability to perform business activities is currently one of the most up-to-date tasks of society in a particular country. Systems of employee preparation to perform these activities are also related to this notion when following the fact that engineering facilities require the same procedures for their operation in safe working conditions in all countries; in reality, their technical properties do not change with the place of their usage.

Risk management systems (RMSs) are distinguished by the same procedures used within quality management systems (QMSs), as well as within environmental management systems (EMSs). The principle is the philosophy of continual improvement when applying the well-known Deming principle, the philosophy known from standards issued by the International Organization for Standardization (ISO). The most frequently used standard for safety management is the Occupational Health and Safety Advisory Services (OHSAS) 18001 standard, established on the same principles as the ISO 9001 standard for QMS or ISO 14001 standard for EMS. The International Labour Organisation has its own version of an occupational health and safety management system, which does not substantially differ from the OHSAS 18001 system.

The reason for defining so-called integrated safety was the complex attitude toward risk management with regard to the important influence of the human factor. Nevertheless, this attitude results from the fact that the same socioeconomic system, most frequently defined as the man–machine–environment system, is analyzed in the field of safety as well as in the field of security. The causal relationship in a negative event occurrence is likewise characterized by the fact that damage in the end is always interconnected with harm. Experience has clearly proven that it is possible to make use of methods shared by safety and security to prevent hazardous situations. The activities of firefighters and the safety of nuclear power plants are parts of both fields (safety and security) and have the same aim, to minimize damages. Both cases observe the fact that it is effective to execute the most efficient regulations within risk minimization in the development stage of objects, machines, additional technologies, or planning suitable logistic systems. Although any other intervention may seem effective and might even prevent damages, it is eventually more costly.

It is obvious from this overview of attitudes within usage of management systems in the field of integrated safety that skills, knowledge, as well as experience gained are prerequisites for successful use of risk management systems as a part of safety culture in any field of social and business activities.

The chapters of this book contain information about specific areas of effective risk management and integrated safety systems, which is a result of more than fifteen years of the author's scientific research as well as realization activities together with the members of his own scientific school and, most of all, with his doctorate students. Some presented in this book were achieved within the solution research projects VEGA 1/017/12 and explore the process of risk management of technical systems to interface safety—safety of technical systems, occupational safety and security, civil security, as well as APW-o337-11. The research shows new and emerging risks of industrial technology within integrated security as a prerequisite for sustainable development and management.

About the Author

Juraj Sinay is head of the Department of Safety and Quality, Faculty of Mechanical Engineering, Technical University in Košice, and the vice-rector for External Relations and Marketing.

In 1995, he earned the degree DrSc with a doctoral dissertation thesis titled "Identification, Control and Predictability of Technical Risks of Transportation Machines and Facilities", and in 1990 he successfully defended his inaugural dissertation thesis, "Beitrag zur Qualifizierung und Quantifizierung von Risiko-Faktorenin der Fördertechnik Dargestellt am Beispiel von Hebezeugen", at Berg University of Wuppertal, Germany, in the field of transportation and handling technologies, which he researched and wrote during a long-term scholarship program within the Friedrich Ebert Foundation.

In 1991, he earned a professor's degree. In 2000 he was named the rector of Technical University in Košice and was the president of the Slovak Rectors Conference from 2002 to 2006.

From 2003 to 2004 he acted as a counsellor for science and research at the President of Slovak Republic's Office. In 2004, he was awarded the first degree Cross of the President of the Federal Republic of Germany for his work in developing cooperation with the Federal Republic of Germany (FRG). In December 2006, he was named a permanent expert of the Government Council for Science and Research in the field of science and technology of engineering sciences. From 2006 to 2008, he was the project leader for the Institutional Evaluation of Slovak colleges by the Association of European Universities. From 2008 to 2010, he was an advisor to the vice-president of the national government and to the Minister of Education for science and research. Prior to 2000, he was a guest professor at Technical University in Miskolc, Hungary, and at the University of Ljublana, Slovenia. Currently, he is a guest professor at the University of Wuppertal and at the University of Applied Sciences, Technology, Business and Design at Wismar in Germany, within the Erasmus program. In 2007, he became a member of the European Academy of Sciences and Arts in Salzburg.

In the field of science and research, Dr Sinay has been working with mechanical dynamics as a source of technical risks, optimization of risk management systems for technical systems and occupational health and safety, problems of control integrated systems, issues of safety and security, and with processes of quality assessment mainly in the field of university education. He has tutored thirteen doctorate students, two of them inaugurated as professors in fields related to safety and one promoted as a senior lecturer in the field of safety of technical systems. He is also an honorary doctor of the University in Wuppertal, Germany; University of Miskolc, Hungary; University of Uzhgorod, Ukraine; and of the College of Mining on Technical University in Ostrava, Czech Republic.

Professor Sinay is currently the president of the Association of Technical Diagnosticians in the Slovak Republic (SR), a member of team of experts of Maschinen- und Systemsicherheit International Social Security Association (IVSS) in Mannheim, Germany, a member of Gesellschaft für Sicherheitswissenschaften VDI in Wuppertal, Germany, and a member of the German-speaking conference of professors of traffic engineering in Europe, located in Hannover, Germany.

He is a member of the editorial board of *Safe Work*, a member of the editorial board, *Human Factors and Ergonomics in Manufacturing*, ISSN 1520-6564 (Wiley, New York, USA), and a member of the editorial board of *DELTA*, scholarly journal of the wood-working faculty at Technical University in Zvolen, Slovakia, ISSN 1337-0863.

Throughout his almost thirty years of scientific work, he has published as an author and co-author a total of ten books (five in the SR and five abroad), two chapters in encyclopaedias by CRC Press in the United States, twenty-three articles in scholarly journals abroad, and nineteen articles in the SR. He took an active part in seventy-three international conferences. He is the author and co-author of forty-seven lectures and publications related to the field of quality of scientific and educational activities within the European high education area, control of science and engineering within the European research area, as well as in the field of technology transfer.

Dr Sinay was the principal investigator of six scientific projects within the VEGA Scientific Agency of Applied Science and Research, the principal investigator of the EU project of TEMPUS, as well as a project manager and investigator of four projects within the Sectorial Operational Programme for Science and Research. He was the leader of twenty-five projects for industrial practice. He is the co-holder of three patents.

1 Safety Culture
Prerequisite for the Development of Modern Society

The notion *culture* comes from Latin, and it means a summary of material and spiritual results of human activity. Therefore, if we talk about *culture safety*, it will mean a summary of all human activities that create conditions for safe work, and life, in the Man–Machine–Environment system (Figure 1.1). A prerequisite for implementing a safety culture is the creation of such conditions, where safety and health protection is a joint task of employers and employees on any level of business management. Accepting this principle has to be conditioned by the fact that health protection has the highest priority in any society and in any sphere of people's lives.

Safety culture in society also includes safety of technical devices. Great achievements have been recently seen in this field. It is not possible to separate workplace safety from safety of technical devices. Both of these fields are managed from the same place, and deliver, together with management of environment protection by a synergic effect, significant economic contributions for the business. Safety is the preferred goal, or Safety First.

Safety is defined as a feature of an object, for example, a machine, technology, or an activity that does not threaten people or the environment. Analyses used to evaluate the complete safety of an object take into account the aspects of safety of technical systems and occupational health and safety. This definition allows us to clearly formulate goals and tasks in the field of occupational safety management. These include all activities related to carrying out and stating the threat extent. When assessing the extent of a threat, as a negative event, it is necessary to state the possibility of its occurrence and assess the extent of possible consequences due to the effect of a negative event, that is, assess the risk. Then it is necessary to assess if the extent of the risk is acceptable. If the risk is higher than acceptable, then it is necessary to execute measures to decrease it or to completely eliminate it. The complex of these activities might be included in the occupational safety management system as a subsystem of risk control, risk management.

Progress in the field of occupational health and safety is shown in Figure 1.2. It is obvious from the picture that occupational safety has gradually become an integrated part of all the strategic activities of a company.

New technologies and machine designs are distinguished by a high level of complexity and get constantly more complex. Nevertheless, things such as their effect

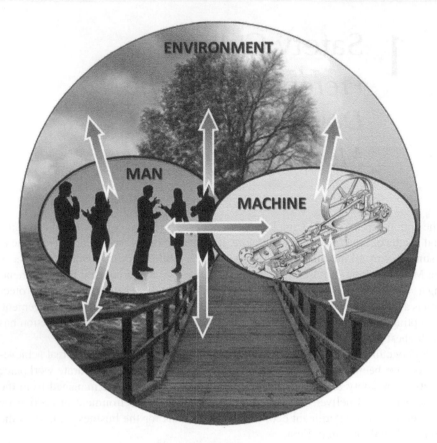

FIGURE 1.1   Man–Machine–Environment system.

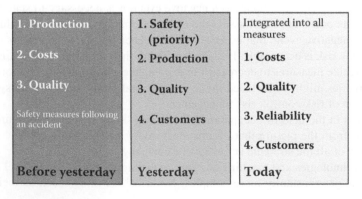

1. Production	1. Safety (priority)	Integrated into all measures
2. Costs	2. Production	1. Costs
3. Quality	3. Quality	2. Quality
Safety measures following an accident	4. Customers	3. Reliability
		4. Customers
Before yesterday	**Yesterday**	**Today**

FIGURE 1.2   Development of safety.

on the environment, ergonomic requirements, and technical solutions to eliminate a breakdown because of human factors are taken into account.

Development trends in technology as part of an intensifying progress of society do not expect to use only new management systems, shared functional interconnection of classic engineering constructions with IT, and therefore new machine designs, new materials (especially nanomaterials) but, naturally, those that have regard for the environment. It often happens during the development and design stages of new types of machines, new solutions that might cause damage during operation, that is, bear risks, come up due to the lack of information about actual operating conditions and due to a missing detailed risk analysis.

The goal of any activity within the design and production stage of machines has to be safe machines, safe control systems and technologies, and safe working procedures in the workplace. These facts push constructors of new machines to execute detailed risk analyses, which are now a requirement of legislative enactments within the European Union and its member states.

Any machine is characterized by the possibility that it may be a threat to operators or third parties. It is necessary to analyze any interface between a machine and man by some of the methods of risk assessment. Scientific research only confirmed that the relevance of analyses results can be expected especially when the identification, quantification, risk assessment, and the choice of methods for risk minimization are not left to the human factor, but are instead carried out using modern procedures from IT. The human factor—man—is not able to process in a short time the extensive amount of information within complex engineering systems or complicated constructions of modern machines.

The basic task of mathematical streams as part of logistic systems is the manipulation of material in a clearly defined space, whereas routes and tools for manipulation might be combined at will. Safe delivery of products or resources in the shortest time possible, to the right place, and at the lowest possible costs must represent the final philosophy. Risks or potential hazards occurring in material streams are functions of various parameters, while the deciding factor is people and their safety.

Development trends in technology and for material manipulation expect the use of new technologies, new machine constructions with high-performance control systems, and new materials. One of the determining factors for their utilization is safety within the Man–Machine–Environment system. It is important to create conditions for safe operation of new machines and technological units as early as in the development stage and consequently at the stage of design work. These procedures are not left to be assessed only for developers and constructors, but they are taken into account by European legislation as well. It is a result of individual countries' priorities in the field of occupational health and safety as well as in safety of engineering systems. It expresses the social development of countries worldwide, whereas the determining criterion is the care of suitable working conditions in any part of the society.

With regard to the increasing level of complexity in sophisticated systems and devices, technology constantly imposes higher requirements for its safety. Therefore, it must be continually assessed considering the influence of the human factor. Nevertheless, the *human factor* means not only the operation of a device, but includes third parties who might be in the operational area of the device, or by chance in the radius of an unwanted condition caused by it. It is obvious that activities within risk management require knowledge of complex relations between technology, work organization, and the human factor. It is even possible to confirm these thoughts by contemplating the number of deaths due to occupational injuries and diseases. In 2007, their numbers climbed up to 2.3 million per year, according to data of the International Labour Organisation. One of the economic consequences was a four percent loss of global gross domestic product (GDP).

A safe machine construction is a prerequisite for executing safe activities within technological processes within the Man–Machine–Environment system. Only safe and functional machines, or complex machine systems, can complete the circle of quality so that the result will be a high-quality product with the potential to sell well.

In specific technological units, it is held that if a machine or device is faulty, then the conflict between safety and quality is even bigger. This means that *weak points* in complex and partial systems must be eliminated to increase their reliability. Increasing reliability means eliminating weak points and, therefore, minimizing risks.

New approaches in the system of risk management (or occupational health and safety management) require every person to be aware of potential risks, both at the workplace and in everyday life. The employer is obliged to identify these risks during the working process, execute measures for their elimination or minimization, and inform employees about any residual risks.

From the technical-analytical point of view, risk assessment leads to the fact that the field of occupational safety and safety of technical systems is directly connected to value criteria, that is, economic analyses.

It is necessary to observe the following to provide these principles:

- Absolute safety does not exist.
- To achieve safety, it is not enough to execute a measure only according to provisions and standards, but also beyond the framework of legal requirements.
- The boundary of safety awareness is not one stable value; it changes subject to the level of technological and cultural level of society and to the degree of knowledge based on the results of scientific research.
- Even detailed analyses and consequently passed respective measures do not guarantee that there is not going to be any injury or any other undesirable event; therefore, preparation to handle breakdowns must be a part of preventive measures as well.
- Employees, machine and device users, and other parties subject to any hazard must be informed about the possibility of negative event occurrence.
- Dangerous situations must be handled by the person who creates them— constructors in their drafts, manufacturers in their products, and employers at work assigned to an employee.
- Conditions must be provided for constant education and mediation of new scientific and research findings in the field of occupational health and safety (OHS) for any group in a society, that is, not only employees and employers, but also anybody outside the working process (third parties).

1.1 TRENDS IN THE DEVELOPMENT OF OCCUPATIONAL HEALTH AND SAFETY

Modern methods of OHS management at work follow these prerequisites:

- A contribution to occupational health and safety for the success of a business, for success of the national economy, as well as for social systems must be clearly stated, although economic reasons cannot be the only reasons. Occupational health and safety is and will remain an important social, human, and ethical duty.
- Occupational health and safety cannot focus only on industrial manufacturing areas including new technologies. It must aim more at all the areas of national economy—and at services in a broader sense. It must find a proper approach to that group of workers that is far from the ordinary employees in industrial technologies.
- Occupational health and safety management systems have to be more flexible to be independent from technical-organizational and social-economic fields of a society's progress.
- Occupational health and safety requires strong partners, for instance systems of social and accident insurance, as well as support in businesses, associations, and professional or corporate organizations.
- The efficiency of measures executed within occupational health and safety has to gradually increase. OHS must become an integral part of business and demanded from partners of employers.
- Occupational health and safety experts must move from the position of simple supervisors to the group of experts who solve problems, or to the group of specialized advisors and to the group of managers. They have to become irreplaceable for businesses as part of a company's top management.
- Legal provisions in the field of occupational health and safety must provide space for effective decision making and new solution suggestions. Attention must be paid to system solutions rather than technical details.
- Occupational health and safety management systems will utilize IT in the field of psychological, mental, and cognitive load on a person.
- Legal provisions for occupational health and safety, as well as for supervision over occupational health and safety, must be effective even in altered conditions, such as outsourcing of production and other activities, establishing an increasing number of small and medium-sized businesses, creation of so-called virtual businesses, and changes in labour markets related to, for instance, an aging labour force and its flexibility.

The following strategic question is very important: In what way does the policy of occupational health and safety have to be formed—if, on one hand, privatization of public institutions is constantly expanding, while on the other hand, deregulation of law in the field of occupational health and safety is applied? Is there going to be a new relationship between state and corporate responsibility?

It is obvious, in the sense of Europe-wide context, that occupational health and safety must progress in accordance with the tendencies of intensive globalization of labour markets both within the European Union and worldwide.

The XIII World Congress on Health and Safety at Work in New Delhi, India, in 1993 stated the following in its conclusions:

- The efficiency of injury prevention must be in constant intensive integration into the management system of a company.
- Occupational health and safety must be substantially, intensively implemented into people's consciousness all around the world. Anyone, whether it is an employer or employee, child or adult, must care about their safety and about the safety of the environment they live in.

The XVIII World Congress on Health and Safety at Work in Seoul, South Korea, only confirmed these prerequisites (Figure 1.3) fifteen years later in 2008, based on the so-called Seoul Declaration, whereas requirements for formulation of effective measures within a modern approach toward OHS management were expanded by the following prerequisites:

- The support of high level occupational health and safety is the responsibility of the whole company and all of their members must contribute to achieve this goal. OHS must be given the highest priority on a national level, while a national culture of preventive occupational health and safety must be created with a permanently sustainable nature.

FIGURE 1.3 World Safety and Health Congress.

- A national culture of preventive occupational health and safety is distinguished by observing the rights for a safe and healthy working environment on all levels. Governments, employers, and employees are actively participating in providing a safe and healthy working environment through the system laid down by a valid legislation.
- Support a constant improvement of occupational health and safety due to systematic measures within safety management, including the development of national politics considering the contents in Part II of the International Labour Organisation's Occupational Health and Safety Convention from 1981 (no. 155).

These comparisons of significant world-known panels in the field of occupational health and safety confirm the constantly high relevancy and importance of risk management and occupational safety as part of any management activities, not only in companies, but in any area of social life.

In order to complete these intentions, it is necessary for managers to become familiar with modern trends in the field of occupational health and safety and to lead by example within their managing activities. Their task is to influence implementing rules of occupational health and safety by providing information, communication, training, and thorough supervision. It is expected that they will become familiar with the principles of occupational health and safety management and that they will be open to opinions of their co-workers. Workplace safety as part of the Man–Machine–Environment system is one of their basic managing activities.

The importance of the mutual relationship between employers and employees as well as the conditions for an effective implementation of safety management systems can be documented by the content of a lecture by Frederick Gregory, captain of spaceships, during the XVII World Congress on Health and Safety at Work in Orlando, Florida, USA, in 2005 (Figure 1.4). In his ideas he followed the safety of space flights and their crashes, which he witnessed as a flight director. He pointed out the fact that important measures for effective prevention are data about the progress of negative processes. Application of the sequence that follows the fact that 100 percent safety does not exist provides for an effective implementation of preventive measures within any technology.

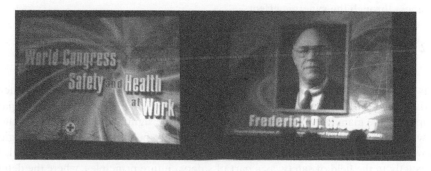

FIGURE 1.4 Keynote speaker at Congress.

This means the following:

- Identify and define the risk.
- Carry out and apply methods for risk minimization.
- Train people in handling residual risk; the aim must be to teach the respective persons how to react properly at the time of a major stressful situation.

An important part of this presentation was to point out the inevitability of applying critical analyses, to openly and retrospectively discuss real risks and their reasons, and consequently, try to simulate the sequence leading to the occurrence of an accident by utilizing modern IT tools. This way it is possible to define the so-called near misses and to look for effective ways to disrupt the causal relation of their occurrence as early as at the modelling stage. This is the only option to effectively develop measures for applying active prevention, which is the basic proposition within the National Aeronautics and Space Administration (NASA). Frederick Gregory concluded his presentation with thoughts on the relationship between an employee and employer who prepare and provide conditions for executing safe work. The point of the basic principle is that employees must rely on their managers and that they will do anything within their power to achieve safe execution of working activities. Above all, it is a question of mutual trust. He compared it with the task of a spaceship captain with other members of the fleet. The basic theory in NASA results from this fact—the right people at right places.

A modern philosophy of activities within OHS results from the application of integrated management systems in businesses, which show the highest level of competitiveness on the market. Activities within occupational health and safety cannot be realized separately anymore, but as a part of management activities, which also include environmental management systems, quality management systems, and therefore every attribute that creates conditions for a high lifestyle level.

Activities within OHS as a part of the Man–Machine–Environment system also include activities related to providing safety of technical devices. Major success has been achieved herein recently. It is impossible to separate workplace safety from safety of technical devices (see Figure 1.5). Both fields are managed from the same place and are a part of safety culture. Together with environment protection management and quality management, it provides significant economic contributions to the business through the synergic effect.

1.2 IMPORTANCE OF MANNER AND FORM OF COMMUNICATION

It is important to create one clear linguistic tool for the realization of management activities within a business or toward society, which is comprehensible for any participant in the manufacturing environment. This issue significantly gains importance nowadays in the age of globalization, when not only people, but also technologies move from one place to another, while a multilinguistic society is being created. It is exactly in the field of safety, as a part of safety culture principles, where the determining factor is the clarity of all the binding and nonbinding legislative provisions,

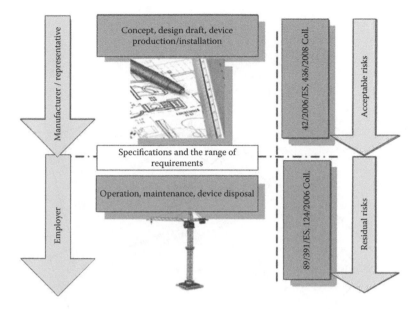

Concept, design draft, device production/installation

Specifications and the range of requirements

Operation, maintenance, device disposal

Manufacturer / representative

Employer

42/2006/ES, 436/2008 Coll.

89/391/ES, 124/2006 Coll.

Acceptable risks

Residual risks

FIGURE 1.5 Relationship between safety of technical devices and OHS.

regardless of what language the participants speak, or where they come from, in order for them to have a clear and single knowledge of every individual determining notion.

Consequent analyses of the language diversity influence must involve the following aspects:

1. Labour markets are becoming international; harms, injuries, and accidents cannot depend on the nationality of every social partner.
2. All of the procedures within risk analyses must be clearly stated and comprehensible for every social partner, even in the international sense.
3. European legislation is implemented in national systems of safety and health protection.
4. Outsourcing of production is constantly intensifying.
5. The effort is made to implement the same attributes of safety culture in specific countries.

Spoken statements allow mediation of feelings on a higher emotional level, create prerequisites for knowledge, and are the most important communication tool between researchers and experts in every area of their expertise, between participants of the manufacturing process on any level in businesses, and even within internationalization of labour markets.

Especially translations of some legislative provisions (Figure 1.6), first and foremost standards of any type, cause potential uncertainty in terms used in specific countries, which might lead to incoherent interpretations by employers and employees, as well as by supervising institutions within internationalization of industrial activities. Therefore, it is important to look for equality in the usage of notions in occupational health and safety management, as well as in safety of technical systems.

FIGURE 1.6   Multilingual legislation.

It is necessary because a negative event will occur at the end of this process, either in the form of a failure, working process interruption, injury, or environmental damage.

Application of the following principles is the outcome of properly handled attributes of safety culture in business:

- OHS must be a part of the business's development strategy.
- Prevention must have priority over anything and be included in the complex of management activities of the business.
- Responsibility within OHS tasks cannot be delegated; it has to be a part of top management.
- The life and health of co-workers have priority over any other decisions in the business.
- An OHS management system must include measures regarding third parties.
- Attention must be paid to minimization of human and material damages.
- OHS must have priority over any other activity within the business.
- Every employee is obliged to observe measures resulting from the OHS management system.
- Create conditions for constant improvement of OHS systems and for their efficiency.
- Provide conditions for lifelong learning of every employee in the business.

The company's management has responsibility for activities in the field of occupational health and safety, for environment protection, safe operation and maintenance, for safety of manufactured products, as well as for offered services in compliance with respective legislative enactments. Top managers in a business are those who create conditions to implement safety culture in the business and carry out every activity for outcome efficiency determination, which comes from implementing OHS management systems as a tool for increasing business prosperity.

Applied principles of OHS management systems have developed into unified theories of safety and prevention. They are utilized above all in the following:

- Construction and production of machines and products
- Design of new technologies and working procedures
- Development of material flow
- Development, production, and distribution of new substances
- Maintenance of technical devices
- Work organization and management system
- Injury and breakdown prevention
- Activities in the field of environment
- Education, briefing, and training of employees

Requirements to carry out analyses of dangers, hazards, and risks imposed both on machine manufacturers and on their users do not regulate what methods to utilize. The choice of procedures and methods is left to everyone's own consideration.

1.3 CHANGES IN OHS MANAGEMENT SYSTEMS RELATED TO BUSINESS COMPETITIVENESS

What has changed from the approach to OHS management in the past, as compared to current requirements of conditions on globalized labour markets related to the creation of prerequisites for business competitiveness?

In the past, OHS management was understood as a process that creates conditions to meet clearly stated legal regulations and standards. Meeting safety provisions was considered as the creation of conditions for safe workplaces.

The current philosophy requires the assessment of all risks at workplaces (regardless of legal requirements) and execution of respective regulations. The prevention principle is actively utilized; this equals *ante factum* procedures—before the act. Safety is understood as the achievement of a certain risk acceptability value, or implementation of a work organization that allows employees to tackle risks that are constantly affecting them and the continuity of the working process.

As one of the management activities, risk management now deals with every aspect of risk. Risks are assessed as a failure potential, which might restrain meeting economic goals of a company. The aim of risk management activities is to quantitatively and qualitatively define these failure potentials and suggest measures that can decrease their values on a level acceptable by every part of the working process. The basis of effective risk management is the development and formulation of a business's safety policy. This policy includes goals of the business's top management in the field of organization and the delegation of tasks and factual competencies of individual organization structures in the business.

Implementation of a suitable OHS management system by employers has an important purpose in reality and currently creates conditions for competitiveness of a business. It is possible to achieve a permanent OHS level increase by creating a proper mechanism that helps a business to run correctly in the field of OHS. That has an important economic purpose since solving issues related to safety and health protection in a broader sense by creating favourable working conditions and relationships results in an optimization of the working process and, hence, a positive economic effect. It also brings a decrease of costs, higher productivity, effectiveness, and quality of work, which means more business prosperity, and therefore, prosperity of the whole society. On the other hand, it also has an important human aspect that mirrors the cultural and social level of a business, country, and of multinational companies as well.

The mobility of foreign capital and management structures positively influences a broader implementation of effective management systems. The requirement to demonstrate reliability through a management system is becoming more relevant in the supply–customer relationship. Larger and prosperous companies focused on acquiring respective certificates as an objective advantage on the commercial market are interested in implementing management systems in general, as well as in OHS management. A number of small and medium-sized businesses are a major area in which the OHS management system can be utilized. It is suitable for them to implement simplified forms of management systems, verified only by an internal or customer's audit. The campaign is mainly focused on the so-called *good neighbour* programs where large companies state conditions of providing a required OHS level as cooperation terms for small companies and subcontractors. This way, it is possible to achieve a unity of working conditions and OHS level even in small companies. The future of OHS management system implementation in small and medium-sized businesses within global market mechanisms lies in this strategy.

Aside from anything else, an OHS management system must observe the following principles:

- OHS creates conditions for constant improvement of management systems and for an increase in their efficiency.
- Attention must be paid to minimization of human and material damages.
- Providing conditions for constant education and mediation of recent scientific research results in the field of OHS for any group in the society, not only for employees and employers, but for anyone outside the working process, including third parties.

The similarity of procedures within environmental risk assessment to procedures within risk management makes the top managers of businesses integrate these activities.

Occupational health and safety is in a breakthrough situation. National legislation observes binding international agreements. In the past, the tasks of traditional safety technicians were related to activities only in one field of expertise. These tendencies are gradually changing. In the future, top functions in the field of complex workplace safety expect one to handle tasks from fields such as OHS, management of technical devices (MTD), environment protection, management of hazardous substances and critical industrial breakdowns, quality management, as well as in the field of explosions and fire protection. Classic activities aimed at retrospective procedures of accident analyses are being replaced by prospective methods focused on the application of modern risk analysis methods at the first stages of a machine's lifespan and development of job vacancies. Therefore, an expert in the field of OHS becomes a generalist within management activities of a company, which then becomes competitive in globalized labour markets.

Quality management is becoming an internal tool of a company's prosperity. Some medium-sized and large businesses have separate departments for quality management, although future progress will lead to a situation where all four systems—OHS management, MTD management, environmental management, and quality management—will be integrated into a single system (see Figure 1.7). It is obvious in conditions of small businesses.

Capability means having specialized knowledge. Capable experts in the field of OHS must be equal partners with their colleagues in the field of planning, development, purchasing, operation and maintenance of machines and devices, as well as in manipulation with hazardous substances. It is therefore natural that such experts must have the highest possible specialized education and must constantly broaden it through a quality lifelong learning program, which is now most commonly and effectively provided by colleges.

FIGURE 1.7   Integrated management system.

1.4 CHANGES IN THE NATURE OF LABOUR MARKETS AND THEIR IMPACT ON OHS

Along with demographic changes in the society, the change in the nature of work and working conditions due to technological innovations and globalization of labour markets presents a major challenge for the field of occupational health and safety and for the development of new forms and methods of worker education in the spirit of the fact that only motivated and healthy employees are a basic condition for competitiveness of companies, and therefore for maintaining and creating new job opportunities.

Globalization is a process defined by an increase of economic relationships and connecting markets of various countries with each other. Changes in working conditions are influenced by the globalization process, since the process was the reason for the creation of globalized worldwide research and education areas.

A part of this process is making the education accessible in fields where it was not before. Globalization requires gender equality, religious tolerance, and unity of legal enactments, along with access to results of scientific research and new technologies. It also means giving a broader access to the labour market, equality of employee positions, their working conditions, and work culture.

Moreover, globalization means cooperation in the fields of science, technology, economics, politics, and social life; it means protection and prevention against natural disasters and terrorism, but also, against negative work effects. Occupational health and safety is a field in which the whole world needs to cooperate. It is no wonder that organizers of the XVII World Congress on Safety and Health at Work in Orlando, Florida, USA, chose globalization as the focus of their main motto: "Prevention in a globalized world—success through partnership." In a broader sense, globalization means that cooperation and partnership enable us to successfully face risks that are becoming common in the globalized world. Therefore, prevention means a common strategy so that occupational health and safety is available for anyone in the world.

Globalizing prevention and creation of international networks presupposes extending the knowledge gained through common activities into every country in the world. Educational activities should be internationalized, that is, one should learn from another. The efficiency of these activities is stipulated by the dialogue and cooperation among all participants in the working process—employees, employers, and governments of respective countries as the lawmakers of valid legislation. Cooperation on a governmental level requires an intensive dialogue among labour, environment, education, economy, and others as well. For instance, Hurricane Katrina represented an underrated prevention; NASA published the information about the intensity of the hurricane nine days before its devastating aftermath. A similar thing happened during a disaster in southern Asia caused by a tsunami, where information about the devastating hit on the coast was made known five days prior to the disaster. A positive example of applying prevention as a part of safety culture is the catastrophe in Fukushima, Japan, in 2011. After the information about the earthquake and a possible tsunami was made public, Japanese people reacted right away and applied measures for risk minimization in full extent. Hence, casualties of this disaster were substantially lower than they could have been.

In a globalized world, the effectiveness of OHS management is stipulated by the following:

- Legislation harmonization
- Definition of motivational criteria; high aim
- Constant system training in one's mother tongue
- Incorporation into integrated management systems
- Presentation of objective information about injury occurrence, that is, negative and positive points of views as well as utilizing current IT tools
- Considering all the circumstances related to the effort of implementing systems of the parent company in subsidiaries established abroad, regardless of national specifications
- Applying lifelong learning; professional capabilities become outdated quickly; consequences of fast progress in IT

Minimization of risks or their complete elimination is an important aspect in applying preventive measures, in addition to their identification in any walk of life.

The World Congresses of OHS always delivers the most progressive results of research, development, application of great work experiences, and challenges for a new approach and cooperation. The field of research is becoming increasingly internationalized. It is not expressed only by an exchange of scientific and specific information, but especially by establishing international scientific teams to solve global problems.

Demographic changes in economically developed countries lead to an aging labour force and hence to the existence of new types of risks that are specific to the aging group of employees. Companies and educational institutions, including colleges and universities, must take into account this development and adjust their educational systems respectively, including the syllabi of study programs, length of educational modules, as well as forms and ways of education.

Changes in working conditions impose new requirements on employees and their qualifications, and hence on new forms of education and the content of the actual educational process. These observations are worth mentioning:

- Labour markets are becoming increasingly multinational. The same methods of risk management are being used within international corporations. Multilingualism is gaining more importance, and is becoming a challenge for linguists. A single European research area is being established.
- Employees must demonstrate their skills in IT, since it is being increasingly utilized. Utilization of IT requires that employees think more in logical, abstract, analytical, and hypothetical ways; mathematical knowledge is becoming more important. Thus, it is necessary to include the fields of IT in various study degrees.
- Professional and social qualifications must constantly improve. More investments are being made in education and in increasing knowledge, especially through a variety of lifelong learning programs.

- Changes in business structures and workplace decentralization require more independence, creativity, initiative and responsibility, and knowledge and skills in communications, cooperation, and team-oriented thinking. In some cases, social capabilities and the ability to work in a team are even more important than other specialized knowledge.
- The work itself is becoming more time and place independent. Employees are required to be more flexible and mobile.
- The population is aging. Relationships between different age groups are changing. The average age of employees is increasing.

1.4.1 New Principles of Occupational Health and Safety

Traditional forms of occupational health and safety presupposed that these were special tasks for a smaller group of experts named by the company's management. These tasks included reacting to problems in the field of occupational health and safety, to injuries and accidents that occurred, and making sure that binding provisions were observed. Paying attention to management systems regarding occupational health and safety is a change in these procedures. A quick change in the current labour market related to gaining and transferring capabilities plays a determining role. It requires a constant application of new forms and methods of occupational health and safety management.

Changing the forms and organization of work and applying new methods of prevention create new requirements for occupational health and safety. Employees that are healthy, motivated, and reasonably charged with tasks guarantee a lasting quality of final products and services. In 2005, the XVII World Congress on OHS in Orlando, Florida, emphasized the importance of OHS within global structures by including a song made especially for this world congress in the opening ceremony. A major Broadway artist, J. Mark McVey, performed the song titled "This Is the Right Moment." Appropriately enough, now is the right time to solve twenty-first-century problems and to look for suitable solutions in partnerships to meet challenges in the field of prevention (Figure 1.8).

In some businesses, methods of prevention within occupational health and safety are included in the corporate philosophy and quality management. These companies realize that employee motivation and satisfaction are both very important business-economic factors. Safety culture is provided by all managers (Figure 1.9) and not only by one responsible worker (Figure 1.10).

Obstacles subject to the nature of work being done are changing due to altered conditions of labour markets and manufacturing technologies. Therefore, new and emerging risks are occurring. Physical loads are less important, whereas mental ones are increasing. Employees often take responsibility for their own safety and health. Provided that they want to handle this, they must have the ability to manage stress, mental loads, criticism, and responsibility, so they are constantly able to motivate themselves and work effectively. Mental resistance requirements are substantially increasing when compared to the ability to rest physical loads. Mental loads are increasing in groups of highly qualified employees. This progress stipulates new requirements that are then imposed on education and training of all employees.

FIGURE 1.8   J. Mark McVey.

Study program syllabi should include fields of study such as psychology, sociology, communication, and so on.

Health requirements now have new importance. Employees want to maintain their health and not be threatened by their working conditions. Health is fundamental for their performance.

Occupational health and safety, as well as ergonomic aspects of the workplace, must be taken into account in business structures and processes in the product/technology design and development stage; then it is not necessary to carry out further

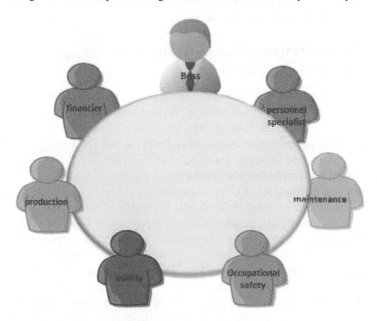

FIGURE 1.9   Staff providing safety culture.

FIGURE 1.10   Human resources management.

improvements or repairs, which are time consuming and require increasing costs. Manufacturers, suppliers, and importers are responsible for providing safe and properly designed machines and devices to the market. They provide users with the information about residual risks and suggest measures for their proper use. Bodies of engineering supervision, as well as occupational safety inspectors, serve as consulting entities for the field of safety of technical devices and support the role of the state in issues of occupational health and safety. Researching and consulting organizations offer services in the field of occupational health and safety management, as well as in the field of safety of technical systems.

1.5   HUMAN RESOURCES MANAGEMENT AS A PART OF SAFETY CULTURE

The current development of management systems significantly affects the integration of OHS into any management process. At the same time, requirements for the complexity of solutions in the field of OHS, especially those of the Man–Machine–Environment system, and for any work-related aspects, are increasing. As a *human factor*, a person (man) is an important element of this system, though it is necessary to perceive them with regard to other elements.

One of the main principles of OHS management is to pay extraordinary attention to the recruitment and training of employees on any level, as well as on the methodology for preparing them and motivating their interest. It is nearly impossible to execute these principles without the support and link to personnel processes.

Human resources (HR) management is a part of the whole organization management. Qualified, capable, and motivated employees are the result of the interaction of personnel processes with other organizational processes. Anything that concerns people in the working process is an HR focus, especially employee recruiting, shaping, and functioning, and utilizing the results of their work, organization, working abilities, behaviour, and relationship to get the work done. Extracurricular activities of employees, which influence the quality of their work within the company, are drawing ever more attention. The main goal of human resources management is achieved by fulfilling two basic tasks focused on the following:

1. Providing an adequate number of employees in the required professional qualification structure and in dynamic accordance with strategic goals of the organization. This does not mean observing only requirements for the number of jobs, but also the variability in the actual specifications for each job.
2. Harmonization of employee's behaviour with the strategic goals of the business. This means an effective integration of employees into the professional life, which presupposes their systematic education and progress as well as effective utilization of their abilities by adequate stimulating tools.

It is this part of management that draws increasingly more attention in developed countries.

Meaningful creation and utilization of human potential presupposes building and developing the strengths and competitive advantages of the organization. All of this is possible based on systematically conceived HR management that manages employees to achieve basic strategic aims and goals of the business.

The effectiveness of HR management has an impact on the effectiveness of the OHS management system. This might be observed by measuring effectiveness with key indexes of performance. Their proper definition can help managers comprehend the need to strengthen feedback between these two fields.

One of the basic requirements of OHS management (according to OHSAS 18 001:2007) includes the definition of OHS policy, which must be appropriate with respect to the nature and extent of safety and health risks of the organization. It must offer a framework for constituting goals and it should reflect the organizational strategy. Therefore, it should take into account adaptation to changing internal and external conditions.

We might include increasing active age of employees into changes of external conditions, which should be taken into account by the organization when constituting the strategy, policy, and goals of OHS. Recent studies dealing with the demographic progress of society have described this trend in detail and pointed out the changes in the age structure of developed countries. They anticipate an increase of people aged 50 and over, which means a higher average age of the labour force, and thus an increase in the retirement age. Governments of specific countries in the world have indeed reacted to the recent population development by similar legislation amendments. For instance, the retirement age in Germany, Iceland, Norway, the United States, and Israel is already 67 years.

Regarding personnel processes, this means the definition of activities that would in fact reflect this part of organizational strategy as well.

1.6 MULTILINGUALISM AS A PART OF SAFETY CULTURE

Linguistics is defined as the study of language and speech. This part of science belongs among the oldest branches, while its research is relevant in any form to social life. Any human activity is related to language (speech), and so to verbal expression of oneself. Speech enables us to express feelings on a high emotional level, creates conditions for environment recognition, and is the most important communication tool between people from any class of social or professional life in general.

The issue of multilingualism gains importance with respect to globalization of labour markets, even though there have been efforts to introduce a single language for work purposes. However, experience shows that usage of only one language will still remain more a wish than a reality.

For this reason, we should note that within communication in the global world of labour markets, the importance of using various languages, their variety, future use, and even the possibility of identifying oneself through a language stipulates activities within applied linguistics. Nevertheless, it is important to include experts from respective fields of language utilization into the specialized staff of projects that are impacted by multilinguistics. This is because only experts of specific fields can accurately formulate the wording of some specific notions that have to be mastered by respective experts. The fields of occupational health and safety and safety of technical devices serve as a great example. In this case an unclear translation of a notion, for example, *hazard*, might eventually cause extensive damage due to an accident or injury.

Nowadays, labour markets are becoming more globalized, and labour mobility is typical for economically worthy investments. Labour mobility is the reason why several business languages should be used within the variety of activities in a company, or why specific communicative materials should be translated into several languages. The European Union is the best example, since all the materials and communication within negotiations here are translated into twenty-three languages (Figures 1.11A and 1.11B).

It is expected that in the future, foreign investors will require potential employees to know their 'corporate language,' especially on higher levels of management. We could mention, for example, a German corporation that acquired a capital share in a Slovak company. They exercise their management rights to use English as the business language. Hence, decisive documents, especially internal provisions and regulations, as well as legislation in the field of occupational health and safety, are translated into several languages.

The field of communication, and therefore effective application of multilingualism principles within OHS management systems and risk management systems, is part of the corporate culture, and as such, of the safety culture of multinational corporations performing in various countries in the world. This happens with respect to some external conditions, including the following:

a. Injuries and accidents have no borders; they are part of international labour markets.
b. Formulation of legislative provisions in the field of safety and risk management must be simple and clear, in order to be comprehensible to every employee in respective countries where the corporation has a subsidiary.
c. Implementation of European legislation in the respective language into legislation of every EU member state.
d. Outsourcing of production to other countries.

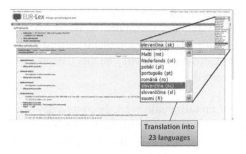

FIGURE 1.11A European Commission webpage.

FIGURE 1.11B Signs.

    e. A unified approach in implementing risk management and safety culture principles into every subsidiary within corporations, regardless of the country in which they are found.

    f. Choice of another communication language as compared to the language of the parent company and the country in which the subsidiary is found, with the aim of providing the explicit meaning of notions within communications on any corporate level.

The following fields can be mentioned as examples of the possible circle where occupational health and safety management systems will practically be applied:

- English uses the notion of *safety* for the safety of machines and engineering systems and occupational health and safety; the notion of *security* refers to civil security.
- In German, there is only a single notion for *safety* (the notion *Bürgerliche Sicherheit* is not used), and that is why, in an effort to unify, the notions of *safety* and *security* are being used more in German civil and industrial practices.

• Identification of dangers, hazards, and risk assessment is executed within risk management systems in Germany. Nonetheless, in English-speaking countries, there is only a single notion that is related to *hazard*, whose translation (either into German or Slovak) means danger, as well as threat, and in some cases even risk.

It could be stated that multilingualism has become a new type of risk, the analyses of which will be the subject of not only linguists, but also working teams of experts from technical fields who use multiple languages within their mutual communication.

BIBLIOGRAPHY

'Challenges of the XVIII. World OHS Congress, Seoul, South Korea, 2008', *Safe Work*, December 2008, pp. 37–40, ISSN 1335-4078.

Pačaiová, H., Sinay, J., and Glatz, J. *Bezpečnosť a riziká technických systémov*, SjF TUKE Košice Edition, Vienala Košice 2009, ISBN 978-80-553-0180-8, 60-30-10.

Sinay, J. 'Riadenie rizika ako súčasť inžinierskej práce', *Acta Mechanica Slovaca*, January 1997, pp. 81–93, ISSN 1335-2393.

Sinay, J. 'Risk assessment and safety management in industry'. in *The Occupational Ergonomics Handbook*, [S.l.]: CRC Press LCC, 1999 S. 1917–1945, ISBN 0849326419.

Sinay, J. 'Risikomanagement-seine Integration in die komplexe Managementsysteme', XV. Weltkongress für Arbeitsschutz-Sektion B4-Neue wirtschaftliche Strukturen und kleine und mittlere Unternehmen, p. 153, Sao Paulo, Brasil, April 1999.

Sinay, J. 'Od BOZP cez kultúru bezpečnosti ku kvalite života', in *International Conference Jakost 2003*, DT Ostrava, May 2003, pp. A3–A6, ISBN 80-02-01558.

Sinay, J. 'Bezpečnosť práce ako parameter konkurencieschopnosti,' in *XVIIth Conference on Current Issues of Occupational Safety*, 2004, NIP Bratislava, pp. 1–8.

Sinay, J. 'Einige Überlegungen zur Selbstverständlichekeit und Notwendigkeit des sicheren Maschinebetriebs in gemeinsamen Europa', Seminar on 25th partnership anniversary of BU Wuppertal and TUKE, dialogue with Prof. Vorathom, Bergische Universität Wuppertal+ TU Košice, April 2007.

Sinay, J., and Majer, I. 'XVIII. World OHS Congress, Seoul, 2008', *Safe Work*, April 2008, pp. 12–14, ISSN 0322-8347.

Sinay, J. 'Sicherheitsforschung und Sicherheitskulturen', Transnationales Netzwerk-Symposium, Bergische Universität Wuppertal, NSR, 29–30, October 2008.

Sinay, J. 'Kultúra bezpečnosti—predpoklad rozvoja modernej spoočnosti', *XXII Conference on Current Issues of OHS*, Štrbské Pleso 2009, pp. 150–155, ISBN 978- 80- 553-0220-1.

Sinay, J. '*Všetci máme spoločný záujem—zdravého človeka a bezpečnú techniku*', *Safe Work*, March 2009, pp. 43–44, ISSN 1335-4078, EPOS Bratislava.

Sinay, J. 'Kultúra bezpečnosti—predpoklad rozvoja modernej spoločnosti', in *Current Issues of Occupational Safety*: 22. International conference: Štrbské Pleso-Vysoké Tatry, 18–20 November 2009, Košice: TU, 2009, pp. 1–6, ISBN 978-80-553-0220-1.

Sinay, J. 'Anforderungen an eine moderne Arbeitsgesellschaft', Arbeitsschutztag Sachsen-Anhalt 2010, Landesarbeitskreis für Arbeitsicherheit und Gesundheitsschutz in Sachsen Anhalt. Otto von Guericke Universität Magdeburg/SRN, 2010.

Sinay, J., and Dufinec, I. 'Management rizík—efektívne vykonávanie podnikateľskej činnosti', in *International Conference Jakost2002, Quality 2002*, DT Ostrava, VŠB-TU Ostrava, May 2002, Ostrava, pp. A22–A26, ISBN 80-02-01494.

Sinay, J., and Majer, I. 'Human factor as a significant aspect in risk prevention', in AHFE International Conference 2008, S.l.: USA Publishing, 2008, p. 5, ISBN 9781606437124.

Sinay, J., and Markulík, J. 'Nové trendy v oblasti manažmentu rizík', in *Ergonómia 2010: Progressive Methods in Ergonomics: Lecture Book*: 24—25 November 2010, Žilina, Žilina: Slovak ergonomical association (SES), 2010, pp. 7–13, ISBN 978-80-970588-6-9.

Sinay, J., Markulík, Š., and Pačaiová, H. 'Kultúra kvality a kultúra bezpečnosti: Podobnosti a rozdielnosti', in *Kvalita - Quality 2011: 20*. International Conference: 17, 18.5.2011, Ostrava, Ostrava: DTO CZ, 2011, pp. A21–A24, ISBN 978-80-02-02300-7.

Sinay, J., Oravec, M., and Pačaiová, H. 'Posúdenie súčasného stavu hodnotenia bezpečnosti technických zariadení', in *Machine Safety Requirements*. Nitra: Agrokomplex, 2004, pp. 31–35.

Sinay, J., Oravec, M., and Pačaiová, H. '*Evaluation of risks as integrated part of modern management systems*', *Acta Mechanica Slovaca* 12, No. 4, 2008, pp. 51–56, ISSN 1335-2393.

Sinay, J., and Pačaiová, H. 'Aplikácia nástrojov na stanovenie bezpečnej úrovne strojných zariadení', *AT&P Journal Plus* 9, no. 10, 2002, pp. 62–64, ISSN 1336-5010.

Sinay, J., and Pačaiová, H. 'Neue Trends bei der Sicherheit und Zuverlässigkeit der Maschinen', in *Conversations in Miskolc 2006*. Miskolc: University of Miskolc, 2006, pp. 71–76, ISBN 9789636617905.

Sinay, J., and Šviderova, K. 'Riadenie ľudských zdrojov v podmienkach systému manažérstva bezpečnosti a ochrany zdravia pri práci', in *Kvalita 2010: Quality 2010*: 19. International Conference, 18–19 May 2010, Ostrava: Lecture book—Ostrava: DTO CZ, 2010, pp. E25–E-30, ISBN 978-80-02-02240-4.

2 Legislative Regulations

Expectations for a Single Approach within Risk Management in the Man–Machine– Environment System

Free movement of goods is the main pillar of the member states market in the European Union (EU) as well as the major moving force of competitiveness and economic growth in the EU. Legislation has helped to set basic requirements on products, provide free circulation of goods, and has contributed to finalizing the single market and its proper functioning. A high level of users' rightful interest protection is set out herein, and especially a high level of occupational health and safety (OHS), which is supposed to be followed. At the same time, it provides businesses with the means of showing equality, which provides the free circulation of goods based on trust in products.

EU directives provide a common framework in the system of accreditation for the purpose of inspecting bodies of equality assessment and bodies of market supervision as well as for the purpose of inspecting products and businesses. The single market functions on the principles of the 'new approach' of stating basic requirements for products according to EU directives and the global approach toward certification and inspection. The market is comprised of twenty-seven EU member states and another three states within the European Economic Area (EEA).

Development trends in engineering assume the usage of new technologies, new machine designs with powerful executive management systems, and new materials. The complexity of modern working and manufacturing activities in successful businesses requires implementation of systematic work arrangements and a management mechanism that provides proper and economically effective operation of the manufacturing process. The level of business management is a condition for fulfilling manufacturing tasks as well as a criterion for competitiveness, a condition for assertion on the market, and a sign of reliability of a business partner following the condition of globalized markets of the EU. It has become obvious even in the new member states' conditions that business partners double-check the level of their suppliers' work arrangement by means of customer audits. A part of this inspection is also the level of manufacturing operation management, quality management, environment, and even occupational health and safety management. If a business plans its future, it tries to implement management systems for specific fields in a transparent way.

As the world constantly changes and develops, working conditions and professional relationships change as well. These changes require a balanced adjustment of job protection. In the 1980s the EU made a significant step in this direction by accepting the strategy of the new approach.

Even the field of occupational health and safety, and the safety of machines and technical systems played a significant role in the ascension negotiations of potential member states with the EU. The ascension process was marked by thorough changes of legal measures which had to be harmonized with the European legal system called *acquis communautaire*. This was a commitment of the candidate countries in the form of conditions to ascend to the EU.

The scholarly public is probably not even aware of how essential the change was in OHS management systems that was introduced by the implementation of the Framework Directive 89/391/EC and Directive 89/392/EC in 1990. Directives have implemented system tools for managing OHS, such as the following: OHS policy, risk management, system of breakdown measures, need for education and inspection, documentation management, integrating employees into solving issues of OHS, and duty to execute risk analyses at every stage of a machine's lifespan. These legislative regulations change prior approaches and procedures in many fields and within OHS management while changing the philosophy in job protection.

The responsibility to execute risk analyses in the workplace and pass the results on to employees results from the Directive 89/391/EC on executing measures to improve occupational health and safety, in article II, section 6, paragraph 2,3 and article II, section 10, paragraph 1, respectively. This regulation changed prior approaches and procedures in many fields, as well as the philosophy of job protection. One of the principles of the new OHS policy is the responsibility to assess risks.

Directive 42/2006/EC on approximation of legal provision of member states related to engineering equipment defines the current European legislation for the field of machine safety. This directive replaces the original version of Directive 89/392/EC, as well as its amendments, Directives 91/368/EC, 93/44/EC, 93/68/EC, and 98/37/EC, which came into force on January 1, 2010.

Directive 42/2006/EC emphasizes raising the level of machine safety, and specifically transfers awareness of threats during operation and maintenance to the design and construction stages of machines. It also emphasizes the creation of measuring points to observe machine conditions by technical diagnostics as early as the stage of projecting the machine. Conditions to prevent malfunctions and then possible accidents are created by determining the real technical conditions; the effect of these conditions would probably have the same consequences on the safety of operation or population. An important aspect is, in fact, a detailed summary of requirements on operational provisions in the mother tongue of the country where the operating technological devices are located. That's how standardized procedures for informing operators about existing threats and residual risks arising out of operation of these devices are created.

Pursuant to Annex I, article I of this directive, a manufacturer of machines or complex engineering equipment is responsible for stating the threats during machine operation, estimate the consequences of a possible injury or harm as well as the possibility of their occurrence, and therefore set and assess risks with the aim to execute measures for their minimization. The responsibility to provide the machine user with any information about residual risks is therefore related to it (bullet 1.7.2.).

These requirements stipulate the activity of machine constructors, constructors of engineering systems, and manufacturers and machine users (including the requirements imposed on their maintenance [bullet 1.6.]). Respective activities must be carried out pursuant to the rules within the complex of risk management.

Taking into account the standards by which machines are constructed, even though their form is not currently binding, provides the constructor with information about how to apply minimal standards when choosing construction elements. This way they will fulfill the criteria of minimal risks and therefore create expectations for the design of safe machines, that is, machines with minimal risks during operation.

Implementation of these procedures as early as at the stage of construction significantly influences expenses. If a wrong safety philosophy has been applied at the stage of machine construction, the user has to reckon with additional expenses in the amount of ten to thirty percent of its purchase price.

Trends in this field stem from the existence of new IT technologies, measuring procedures and monitoring devices, technologies, and materials, which one can deem as new types of hazards and include them in the process of risk assessment at the draft stage.

In the future, we can expect further developments in the field of requirements for safety of especially complex mechanical devices due to the application of a form of *integrated approach* coming from legislative requirements for safety and health protection of employees and affected populations during risk assessment.

We might conclude the requirements of Directive 42/2006/EC for its implementation into real processes as follows:

- Harmonized, efficient, and single-market supervision.
- Specification of clear, distinctive regulations on the level of European Communities for 'risky' equipment, where there are not enough harmonized standards for its implementation.
- Construction of safe mechanical devices must employ the newest know-how of science and technology while taking into account economic requirements.
- Stating a single procedure for partially complete mechanical devices (even if not in their whole extent) in order to provide their free movement.
- An important aspect is the directive's requirement to create harmonized standards related to prevention of risks from construction of devices with the aim to help manufacturers demonstrate equality.
- Stating new procedures for equality assessment related to the extent of the threat, that is, to carry out specific procedures for each category of mechanical devices pursuant to the Council Decision 93/465/EC.
- Unanimous statement of full responsibility of the manufacturer for confirming equality of mechanical devices.
- The CE mark is the only one to provide for the requirement of equality with this directive, and it is inevitable to implement it as a whole.

- The manufacturers, or their representatives, must provide for the execution of risk assessment for mechanical devices that they plan to introduce to the market. They have to state basic requirements for safety and health protection related only to that device, which requires respective measures to be accepted.
- Member states should acquire proper tools, and there are sanctions for violating the requirements of this directive.

Devices subject to the requirements of this directive are the following:

a. Mechanical devices
b. Replaceable attachments
c. Safety parts
d. Lifting equipment
e. Chains, ropes, and arresting gear
f. Detachable devices for mechanical transfer
g. Partially completed mechanical devices

A new approach is to define conditions for equality assessment of the so-called partially completed mechanical device (Art. 13 of Directive) "that presents a kit which is almost a mechanical device, but unable to fulfil a 'certain purpose alone'." A driving system is a partially completed mechanical device. A partially completed mechanical device is meant to be built into another mechanical device or partially completed mechanical device or as a mount between them, which creates a mechanical device.

Conditions for introducing the product to the market are stated as specific requirements on safety, and especially on the so-called integrated safety, which requires focus on defining problems related to safety during the whole life cycle of a device, as well as considering conditions of its incorrect use. This means that the level and extent of regulations must account not only for the machine's features, but also for the conditions of its use. The aim of such accepted regulations must be the elimination of any risk during the whole expected lifespan of a machine.

The European Directive 89/392/EC (now 42/2006/EC) on machines and principles of introducing products to markets was adapted in the Slovak legislation in 1999. General engineering requirements for products and their equality indication set the basic principles valid for any product.

This directive regulates the following:

a. How engineering requirements are imposed on products that could threaten health, safety, or property of a person, or human environment (abbreviated as 'rightful interest')
b. The rights and duties of a legal entity entitled to activities according to the law, which is related to creating, approving, and publishing Slovak engineering standards
c. Equality assessment procedures for products with engineering requirements

d. The rights and responsibilities of entrepreneurs and other legal entities set under a special law and entitled to activities related to equality assessment according to this law

e. The rights and duties of entrepreneurs to manufacture, import, or introduce products to markets

f. The scope of a central state administration body and other bodies of state administration on the section of technical normalization and equality assessment

g. Law enforcement supervision, including imposition of fines

2.1 MAINTENANCE AND SAFETY OF MACHINES

New approaches in the field of machine safety also cause changes in the conditions for their maintenance. The ISO/TS 16949 Standard for the automobile industry and its suppliers mainly requires the implementation of predictive maintenance to increase the quality level. The directive on machine safety states a requirement to be followed at the stage of construction and design: "in case of automated devices, and if deemed necessary, other devices, a connecting device must be provided for mounting [a] failure diagnostics device."

Maintenance management as one of the possibilities to effectively prevent engineering risks has become a progressive field through cost cutting, but mainly by minimizing risks in machine operations. As early as the design stage, the support for technical device care is one of the basic necessities for implementing progressive maintenance approaches based on the minimization of negative effects on safety and environment, as well as on optimization of costs and their maintenance (Reliability Centered Maintenance [RCM], Total Productive Maintenance [TPM], and Rick-Based Inspection [RBI]). However, the reality of these results depends on the quality and compliance with this procedure for implementing the following steps in a company:

a. Goal assignment: defining the aims of a company's management in their compliance with the goals of maintenance management; alternatively, the assignment of additional goals resulting from legislative requirements (directives, laws, standards, etc.)

b. Device data analyses: actuality, manner of their collection and records, form of the record, level of interrelatedness with other data (e.g. spare parts stock, suppliers, external services, etc.)

c. The extent and support of implementation: stating the step sequence, assignment of responsibilities for particular phases of implementation (time, finances, human resources), types and extent of training

d. Specification of a suitable tool and form of output: for example, software support in the form of a new application or making use of Microsoft Excel output, maintenance schedules, statistical indexes, maintenance performance indexes (key performance indicators [KPIs])

e. Feedback, regular meetings of management led by the project's sponsor to eliminate any undesirable procedures, specifying further actions, audits, benchmarking, etc.

2.1.1   Place of Technical Diagnostics during the Risk Analysis Process

The basic requirement imposed for mechanical systems, technologies, and machines stems from the assumption that a product has to be safe during any of the stages of its technical life, that is, even during its operation, including maintenance and repairs. These features must be provided within the Man–Machine–Environment system as well as when taking into account their mutual influence. It is in fact this system that is subsequently the object of executing safety analyses.

Modern maintenance methods as part of risk minimization for technical devices and complex technologies are distinguished by an interdisciplinary nature. They include the methods of observing the actual conditions of a machine, methods of technical diagnostics (Figure 2.1), making use of applications for mathematical statistics, and taking into account the human factor.

Technical diagnostics has noticed a constantly increasing interest in application of its results within newly developed methods for technical risk minimization as a part of preventive measures to achieve flawless and safe operation of technical systems, complex technological units, and particular machines. Making use of technical diagnostics is only one of the basic tools for effective maintenance management and elimination of undesirable conditions of technological devices, affecting production or other aspects that influence the company's prosperity.

Technical diagnostics is irreplaceable in any field of modern society development including:

- Human resources and education
- IT society
- Business environment
- Science, research
- Innovative technologies

Innovative technologies include their safe and reliable operation, within the effective business environment, which includes safety and reliability of technical devices.

Utilization of technical diagnostics methods, when taking into account the reliability of results, is only one of the basic tools for effective maintenance management and elimination of undesirable conditions (hazards) as a result of technical or human factors affecting the safety of the Man–Machine–Environment system.

Defining specific hazards is a condition for the causal dependence of negative events, for example, malfunction, accident, injury, to be interrupted by technical solutions.

2.2   CONTENT OF THE USER'S MANUAL

A detailed user's manual for a machine, and its structure and form, are important in making the user aware of the safety risks associated with machine operation. Several negative experiences generated pressure for increasing the emphasis on the quality of manipulating and implementing this manual, as defined in the 42/2006/ES Directive.

Every user's manual has to contain at least the following:

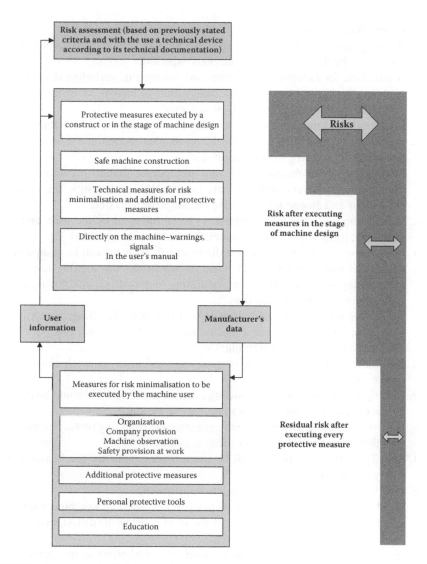

FIGURE 2.1 Methods of risk minimizing.

a. The manufacturer's and its representative's business name and complete addresses
b. The designation of the mechanical device, as stated on the mechanical device itself
c. An equality declaration, or a document, that includes the content of an equality declaration, while stating the details of the mechanical device
d. A general description of the mechanical device
e. Drawings, schemes, descriptions, and explanations that are necessary for the use, maintenance, repair, and inspection of the mechanical device
f. Workplace description(s) where the operation will probably take place

g. Description of the mechanical device's future utilization
h. Warnings about prohibited uses of a mechanical device; these prohibited uses might be defined based on previous experience
i. A guideline for mounting, installing, and connecting, including drawings, schemes, fastening tools, and the designation of chassis or installations, where the mechanical device will be mounted
j. Installation and mounting guide with the aim to reduce noise or vibrations
k. Guidelines for launching and using the mechanical device, and if necessary, guidelines for training the operation staff
l. Information about residual risks that still persist though regulations on implementing a safety point of view into the phase of design; the safety regulations and amending protective measures were passed
m. Information about protective measures to be followed by the user, and if necessary, including personal protective tools that are provided
n. Basic features of tools with which the mechanical device can be equipped
o. Conditions under which a mechanical device fulfills the stability requirement during operation, transport, mounting, demounting, shutdown, testing, or unpredictable breakdowns
p. Guidelines for providing safe execution of transport, handling, or storage, with regard to the weight of a mechanical device and its various parts in case they are transported individually
q. Procedures after breakdowns or injuries; guidelines on unblocking the device safely
r. Description of activities for adjusting and maintenance that should be done by the user, and the measures for regular maintenance
s. Guidelines for safe adjustment and maintenance, including protective measures that should be followed for these activities
t. Specifying spare parts that will be used, if they influence the health and safety of operators
u. Information on airborne noise emissions including:
   • A noise emission level at a workplace as measured by weighted filter A, if the level exceeds 70 dB(A); if the level is below 70 dB(A), this must also be stated
   • A maximum momentary noise level at a workplace as measured by weighted filter C, if the level exceeds 63 pa (130 dB with the reference point of 20 µPa)
   • A level of machinery acoustic power as measured by weighted filter A, if the noise levels at a workplace, determined by weighted filter A, does not exceed 80 dB(A)
v. Information for operators and threatened persons about emitted radiation, in cases where it is possible that a mechanical device will emit nonionising radiation that is harmful to people, especially those with active or nonactive medical implants

The user's manual in the official language of a particular country of the EU must be included in each delivery.

2.3 ILO-OHS 2001 MANAGEMENT SYSTEM

Several known procedures or guidelines, such as the International Labour Organisation (ILO) Directive OHS 2001 and OHSA 18001, are available to help execute activities of occupational health and safety management.

It is definitely important to focus on building a system of health protection as a basic element of constructing a system of safety culture and its implementation in management procedures within a company if we take into consideration the information that four percent of gross domestic product (GDP) of most countries is lost due to injuries (ILO data).

The International Organization for Standardization (ISO) has been dealing with ideas to introduce an ISO standard for occupational health and safety management systems for a long time now. The main idea was to amend the existing ISO 9000 standards quality management system and the ISO 14000 and environmental management systems, with a compatible standard for occupational health and safety management which initially had the class of 18,000 reserved. A decision was passed in Geneva in 1996 for such a standard to be created which would follow the philosophy of ISO 9000 and ISO 14000 standards. However, the draft already considered the fact that it would not be an ISO standard, but a directive of the ILO. In 1999, the ISO tried to form a working group to create a standard in which the ILO would take part as well. Nonetheless, the ILO committee executed a survey in which the member states expressed their approach toward this standard, due to disunity in the opinions of specific ILO member states to applying the ISO standard for OHS management. Economically developed countries were in most cases against implementing such standard for OHS management. The reason was a concern about a commercial approach when applying the standard. Certification within ISO 900 and ISO 14000 presented a similar situation. Furthermore, the argument that OHS is included in legally binding legislation in most countries and that it is in the country's interest to apply OHS management systems (subject to laws of each country) in specific organizations seemed to be yet another sticking point. One of the basic goals of this document would be its flexibility in application in both large and small companies. Specific organizations would then modify it according to the type of industry where they are effective.

Based on the results of the survey, the ILO decided to create a working group that will draft a technical guideline for the structure of occupational health and safety management systems that will be offered to every member state for its application under conditions of specific countries without being considered as binding. It will also be compatible with the ISO 9000 and ISO 14000 systems. It is possible, based on this guideline, for each country to adjust the basic text of the guideline to the conditions of their management systems while taking into account their national legislation, that is, so they can create their own national occupational health and safety policy as a part of the nation's safety culture. Furthermore, it comes from the assumption that every country will create its own national program of occupational health and safety, which will be executed under supervision of a 'competent institution,' in this case the National Labour Inspection of the Slovak Republic. The goal of this program is to regularly check the level of its implementation, support

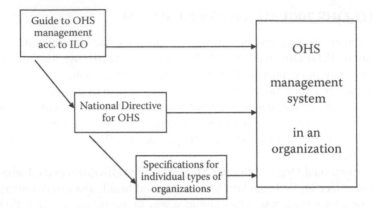

FIGURE 2.2   Directive scheme for OHS management systems.

cooperation between social partners, offer counselling, provide publicity of specific activities within the program, evaluate the effect of state policy, and so on.

The philosophy of the ILO guidelines for occupational health and safety management systems is seen in Figure 2.2.

The commission carried out a draft of the directive and presented it during an ILO meeting at the A+A conference and exposition in Düsseldorf on May 14, 2001. Some of the characteristic elements of the directive might be summarized in three basic points:

- The directive does not mention companies, but organizations.
- The content is not binding, that is, *can* or *might* is used instead of *must*.
- The system does not consider any certification.

Even though the authors tried to create a directive that would not be marked by other prior directives already used in some countries, we might state that its content is substantially similar to the Occupational Health and Safety Advisory Services (OHSAS) 18001 of 1999.

2.4   STRUCTURE OF THE ILO DIRECTIVE

The actual text of the directive assumes that occupational health and safety, including providing supervision to make sure that its principles are followed, are in the employer's full competence.

The occupational health and safety management system includes these activities (Figure 2.3):

- Occupational health and safety policy
- Organizing
- Planning and implementation
- Evaluation
- Improvement measures

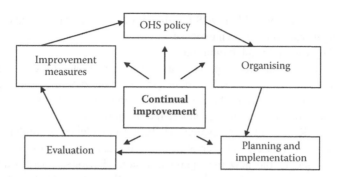

FIGURE 2.3   OHS system management.

A. Occupational health and safety policy

The employer, together with the employees, carries out a basic philosophy of executing an occupational health and safety policy for their
organization that is 'hand-tailored for themselves.' It also follows that
the employer bears responsibility for OHS. The form of each point
and goals must be clearly stated and has to provide for their possible
amendment according to the most recent know-how. Every employee
of the organization has to be familiar with it. A clear specification
of competencies in the management system is another important part.
The OHS policy must come from valid legislation of a particular country, or it must motivate lawmakers to draw up compatible legislation.
If there is already a quality management system, or even an environmental management system within the organization, then the OHS
management system should be implemented so that it is possible to
integrate all three systems into a single one (especially in small and
medium-sized businesses).

Participation of employees in the execution of the OHS policy is a very
important detail. It is necessary for employees to participate in creating
the structure of the OHS policy, to be informed in detail about its goals,
and to be constantly trained.

B. Organization

The employer is responsible for executing the OHS management system
in the organization. They have to clearly state the responsibilities of
every employee and their responsibility on particular management levels. Furthermore, the employer bears the responsibility for identification, evaluation, and management of risks. The employer will prepare
comprehensible documentation that is part of any activity within the
management system. The communication system plays an important
role, both inside the organization and in relation to external partners
who participate in the occupational health and safety management system (e.g. state supervision). It is necessary to name a manager within
the organization's top management who has full control over the field of
occupational health and safety. A system of continuous education and

training in the field of OHS for all employees must be carried out in the organization at its own expense.

C. Planning and implementation

The main task at this stage is to assess and evaluate the existing occupational health and safety management system within the organization. Attention must be directed first and foremost toward fulfilling requirements of valid legislation and the ability to identify and evaluate risks. Results of these evaluations have to be clearly formulated, since they will consequently create a basis for defining requirements to be included in the new management system. Periods to fulfill activities within executing particular measures will be stated at the same time. Both the deadline and the content of each activity must be objectively formulated in order to be fulfilled efficiently by individual bodies of the organization or even by individual employees. Another important role is played by the definition of clearly formulated measurable OHS goals.

The centrepiece of the OHS management system is aimed at risk prevention. In respect to this fact, it is necessary to direct attention to the continual process in the system of risk management as defined for example, by standard STN EN ISO 14121-1 and to apply the hierarchy of risk management possibility.

Before the employer attempts to execute changes in the existing OHS management system, it is necessary to objectively evaluate if they are executable in the organization and if the employees are ready for them. If not, it is necessary to execute an effective form of training. In this respect it is very important to pass conclusions in the sense that every purchased operation for a particular organization must meet the same conditions of the OHS management system which is used by the organization purchasing these operations.

D. Evaluation

An essential part of the evaluation stage is the actual monitoring of every measure applied within the OHS management system in an organization. Furthermore, it defines the responsibility for its execution on every management level. The aim of monitoring is to acquire information on fulfilling goals of the OHS policy with emphasis on risk management. The monitoring has to be first and foremost proactive, that is, it has to be executed regardless of whether an undesirable event happened or not, and its results must be simply and clearly documented. A part of the evaluation process is the reactive activity monitoring within the OHS management as well, for example, by investigating occupational injuries and diseases. Special attention must be paid to formulating conclusions and use them to motivate employees. Every employee in the organization has to be informed about the results of monitoring. Results of internal investigations are as important as external investigations, for example, by labour inspectors.

Auditing is another important part of the process of evaluation. It is necessary to state the audit's program with the competencies and content of the auditors' activities. The audit must include, first and foremost, information about the content of the OHS policy, employee participation, responsibility delegation, competencies, and qualification expectations, documentation in the OHS management system, and about the monitoring activity within the risk management system. It is necessary to emphasize the recruitment of auditors who might be internal or external experts, but they must be independent of the activity that is being audited. The audit results must evaluate the efficiency of the management system and state whether the OHS management system contributes to its continual improvement. The audit results need to be discussed with all employees of the audited operation.

E. Improvement measures

A proposal for the execution of particular measures, based on the results of observing the efficiency of the OHS management system, on the audit's results, and on regular checks, will be carried out. These measures will then be applied in the form of preventive activities or immediate correction measures. Measures that provide continuous improvement of particular elements of the OHS management system, as well as the system as a whole, will have the highest priority. If possible, it is recommended to compare the structure of the OHS management system with the results achieved in other organizations in similar fields.

2.5 OCCUPATIONAL HEALTH AND SAFETY MANAGEMENT SYSTEM OHSAS 18001

The most widely used document in the field of OHS, the OHSAS 18001 (and its interpretation, OHSAS 18002), also provides guidelines for implementing the OHS management system. This standard originated from the initiative of fourteen certification organizations from all around the world as a response to the ISO for not passing a standard. It is not a regular ISO standard, as in quality or environmental management systems. However, most of the companies have been using it as a part of their integrated management systems.

The OHSAS 18001 (OHS) management system was introduced on April 15, 1999. Twelve renowned institutions participated in its development, from the field of occupational health and safety and from fields of certifications, audits, and normalization from seven countries.

The OHSAS directive was created to be compatible with international standards ISO 9000 and ISO 14000, and aimed at providing their integration into common management systems in organizations that pursue it.

The directive's structure contains requirements imposed on the occupational health and safety management system that will allow a particular organization to provide risk management, and improve their economic results. Activities resulting from the directive might be used for any kind of organization whose aim is to eliminate or minimize risks, improve the OHS level, or certify their management system.

If we compare individual elements of management systems according to ILO (Figure 2.4) and according to OHSAS 18001, we might state that these systems are

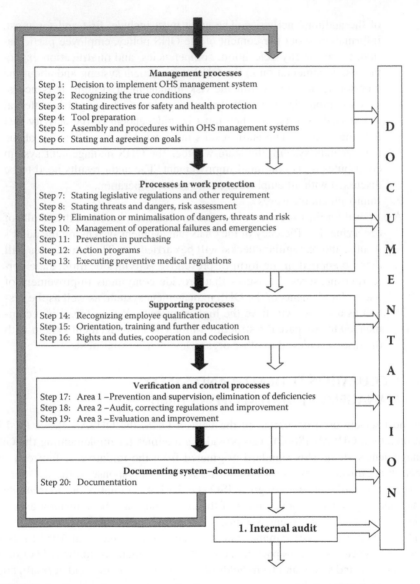

FIGURE 2.4   Occupational health and safety management system in twenty steps.

identical in most ways. The only difference, even though only according to its formal arrangement, is the fact that the OHSAS directive requires that top management deal with the results after execution of improvement measures.

One part of the directive, with respect to the creation of an integrated management system, is Annex No. 1, where the mutual relations between the OHS management system, environmental management system, and quality management system are mentioned.

Thus, it can be stated that the OHS user's manual according to ILO follows the OHSAS 180001 directive.

2.6 OCCUPATIONAL HEALTH AND SAFETY MANAGEMENT SYSTEM OHRIS

The Bavarian State Ministry of Labour, Social Affairs, Family, and Women developed the OHS management system OHRIS, in cooperation with industrial companies in 2001. This system not only includes the OHS management system, but the management system for safety of technical devices. This system was tested at the first stage in some of the larger companies. It is known that production is indeed run mostly in small and medium-sized businesses, which report the most occupational injuries. Therefore, one of the main goals of the OHRIS authors was to amend the system for the needs of these companies. The system structure is designed to enable using it within integrated management systems.

The OHRIS management system's structure is divided into 20 steps. According to Figure 2.4, it is obvious that all of the elements used in the management systems according to ILO and OHSAS 18001 have been included. However, the difference lies in the fact that particular elements are specified in more detail. Therefore, it is possible to use them flexibly for various types of companies in several industry areas. This management system also mentions in its Annexes 3 and 4 the possibilities of integrating individual management systems into a single one. Concrete examples of applying each element of the management system are definitely more present in the text of the OHRIS directive than they were in the management systems already in use. This substantially shortens the time of its implementation in a company.

2.7 A CHANGE IN THE PHILOSOPHY OF OHS MANAGEMENT SYSTEMS RELATED TO ACCESSION OF NEW COUNTRIES TO THE EU

So what triggered a change in the approach to OHS management in the past, as opposed to current requirements in the conditions of globalized labour markets? OHS management was understood in the past as a process that creates conditions to fulfill clearly stated regulations, legal provisions, and standards. Fulfilling safety provisions was considered as the origination of safety at a workplace.

As one of the management activities, risk management nowadays deals with every aspect of risk. Risks are assessed as breakdown potentials that might obstruct businesses in achieving their economic goals. The aim of risk management is to qualitatively and quantitatively define these potentials and suggest the measures that might decrease their values on a level acceptable by every part of the working process. The basis of an effective risk management is the development and formulation of the company's risk policy. This includes goals of the top management in the field of organization, task delegation, and subject-matter competencies of specific organizational structures of the company.

Implementing a suitable OHS management system for employers in the Slovak Republic (SR) has an important justification in reality. It is possible to achieve a permanent increase of the OHS level by creating an appropriate mechanism that provides for proper activity of a business in the field of OHS. This indeed has significant economic importance, since solving issues regarding occupational health and safety, and in a broader sense regarding creation of favourable working conditions and working relations, brings an optimized working process and positive economic effect. It creates conditions to cut costs, improve productivity, and enhance efficiency and quality of work, which means higher company prosperity and, therefore, higher society prosperity as well. On the other hand, it has an important human aspect, too, which presents the cultural and social level of a business, country, or even multinational companies.

Integration of the OHS management system in main management systems has a constantly intensifying use in a number of small and medium-sized businesses. A strong campaign in countries of the European Union is focused on this group of businesses, considering the fact that they create the largest value in the gross domestic product in any economically developed country. It is appropriate for such businesses to implement simplified versions of management systems, verified only by an internal or customer audit. Attention is paid to procedures when large companies state conditions for systematic assurance of a required OHS level as the condition for cooperation with small companies and subcontractors. A unification of working conditions and OHS level, even in small companies, can be achieved in this way. The future of asserting the OHS management system in small and medium-sized businesses not only in Slovakia, but in new EU countries, lies in this kind of strategy.

Occupational health and safety finds itself in a breakthrough condition. Tasks of a traditional safety technician in the past were connected to activities in only one field. However, these tendencies are gradually changing. Peak functions in the field of complex workplace safety expect in the future to cope with tasks in the field of OHS, safety of mechanical devices (SMD), environmental protection, management of hazardous substances and critical industrial accidents, quality management, as well as in the field of explosion and fire protection. Classic activities aimed at retrospective accident analysis methods are replaced by prospective methods aimed at application of modern risk analysis methods at the initial stages of a machine's technical life and projection of job positions.

Quality management has become an internal tool of company prosperity. Some medium-sized and large companies have separate departments for quality management, although development in the future might lead to a situation where all three systems—OHS and SMD management, environmental management, and quality management—will be integrated into a single system. It is already happening in small companies.

Capability means having specific knowledge. Capable experts in OHS must be equal partners with their colleagues from the fields of planning, developing, purchasing, operation and maintenance of machines and devices, and manipulation of hazardous substances.

BIBLIOGRAPHY

Pačaiová, H., Sinay, J., and Glatz, J. *Bezpečnosť a riziká technických systémov* SjF TUKE Košice Edition, Vienala Košice 2009, ISBN 978-80-553-0180-8.

Sinay, J. 'International Labour Organisation Directive on occupational health and safety management: Its impact in integrated management systems', XIV. Conference Current Issues of Occupational Safety, VVÚBP Bratislava, October 2001, pp. 63–71.

Sinay, J. 'Einige Überlegungen zur Selbstverständlichekeit und Notwendigkeit des sicheren Maschienebetriebs in gemeinsamen', Europa Seminar of the 25 Partnership anniversary of BU Wuppertal and TUKE, dialogue with Professor Vorathom, April 2007, Bergische Universität Wuppertal+ TU Košice.

Sinay, J., and Laboš, J. 'Zákon Č. 264/1999 Z.z. NR SR a jeho aplikácia pri konštruovaní zdvíhacích strojov', in *Zdvíhacie zariadenia v teórii a v praxi*, Košice: TU-SjF, 2000, pp. 16–22, ISBN 8088896134.

Sinay, J., Oravec, M., and Pačaiová, H. 'Nové požiadavky Európskej smernice na bezpečnosť strojov a jej dopady' (New requirements of European directive on equipment's safety and its consequences), in *Defektoskopie 2007*, Brno: VUT, 2007, pp. 211-216, ISBN 9788021435049.

Sinay, J. 'EU Directives 89/391/EC and 89/392/EC and their place in the integrated quality management system', VI. International Conference Trial and Certification of MASM, SSK, Slovak Quality Association, Žilina, 1998 under the working title 'Providing consumer protection by safe and ecological goods', pp. 42–49.

3 Risk Management as a Part of Integrated Systems for Business Management

The World Health Organization (WHO) interprets *quality of life* as the links between quality of life, material conditions, culture, and ideology as generally known beliefs. The mutual influence of these specific aspects itself defines the quality of life standards. Therefore, it is too difficult to define it generally. It resembles the definition of a good quality car. The collocation *quality of life* is directly/indirectly related to human life.

Quality of life might be understood as a multidisciplinary phenomenon. We may use various indicators for its description. Quality of life is not a recent phenomenon though; its roots go way back into the past. At first, it was perceived at two levels: a spiritual level (religious) and a philosophical one. Within the quality of life analyses we might state that every person in this world lives and acts in order to move on to a qualitatively higher level. Education and studying, exchange of an older car for a newer one, or changing our computer for a newer, faster one testify to this fact.

Detailed analyses of the quality of life date back to the beginning of the nineteenth century. At that time, the quality of life was perceived through social indicators, whereas quality of life in the middle of the twentieth century was regarded as wealth, profitability, and consumer lifestyle. Nowadays, the interest to investigate nonmaterial values has been increasing. Hence, quality of life is becoming a complex, multidisciplinary notion.

Quality of life might be measured by various indicators. In reality, there are a number of values for its measurement. It has been shown that a constant increase of the number of parameters does not contribute to objectification of these measurements, but rather the other way around. Basically, we may define nine main areas that are considered within the quality of life assessment:

1. Material wealth (economic prosperity)
2. Health (available health care and its level)
3. Political stability and safety
4. Family life
5. Social life
6. Climate and geography (the type of climate is taken into consideration)
7. Job stability (including occupational safety and unemployment rate)

8. Political freedom
9. Gender equality

It is rather ironic, however, that while there is a general agreement as to what quality of life is *not*, there has not been an agreement yet on what quality of life truly *is*. The reason for this lack of consensus is the comprehension of life itself. Nowadays, democracy and its negative message that almost 'anything is allowed' echoes throughout the developed countries in the world. There are still opinions that differ when defining life as it is. Naturally, that results in the inability to find clear value criteria for quality of life, since everyone judges quality of life according to their own criteria. It is similar with product quality, because it is not clearly stated what is of good quality and what is not.

Values are the decisive aspects that are observed in quality of life assessment. These are the aspects that guide and enrich people who believe in them. Values change in the long run; they are not defined in a matter of time. Different perception of values themselves causes a different perception of quality of life.

When we buy a product it is important whether it will meet our requirements, imposed on it by the customer's general expectations. This principle is obvious from the definition of quality. Is it then possible to isolate quality from safety or even set one over the other? Everyone who buys a product (let it be anyone) will put an emphasis on performance as well as safety parameters. It is unacceptable, for instance, for a car to be fast and efficient with an attractive design without standard safety equipment. Naturally, safety has an important place when choosing a product. Requirements for minimal risks are a part of a complex of requirements imposed on products as well as on manufacturing technologies. Therefore, it is not right to give a higher priority to quality or safety.

Recently, the topic of *safe life* as a part of civil safety has been discussed in relation to quality of life. However, it is not anything new in this era of intense progress in science and technologies, especially in IT. These discussions are held above all about what should have priority, whether quality or safety. It is possible that experts will not come to a clear conclusion. Quality and safety are two things that are closely related. A decreasing level of safety in manufacturing technologies means a decreasing marketability of products, and hence their quality, from the customer's point of view. A good life is a life in conditions of minimal risk in every area of one's active life.

3.1 SAFETY CULTURE AND QUALITY CULTURE

Recently, the notion of *culture* has been often used during the application of new management systems. This word has its origins in Latin and means the summary of material and spiritual results of human activity. Thus, if we talk about *safety culture*, it means the summary all human activities that create conditions for safe work and life in the Man–Machine–Environment system. As for *quality culture*, it means the creation of conditions for execution of any activity within the quality management systems in order for them to result in a final product that meets the requirements of interested parties.

The creation of conditions in which safety and health protection are understood as a joint task of employers and employees on any level of organization management is a prerequisite for implementing safety culture. Likewise, the final quality of a product is the summary of qualities of specific activities that are part in its production. Accepting this principle has to be conditioned by the fact that health protection is the prioritized fact in every organization and in any area of human life. Let this be held—safety is the prioritized goal or safety first.

The International Organization for Standardization (ISO) 9001 standard defines quality as a degree of its own specifications that meets the customer's requirements. We might state with regard to this subject that safety is one of the elements of quality. In order to increase the quality of machines and therefore their safety and reliability, it is necessary to apply methods of quality management systems in accordance with up-to-date scientific and technological know-how. Reliability is obviously one of the features of a safe product. It is the fulfillment of its working/user properties with respect to its stated requirements as defined in its technical conditions or manuals. It is therefore only natural that there are a number of common aspects and mutual links among quality management systems, safety of technical systems management, and occupational health and safety management.

3.1.1 RELATIONSHIP BETWEEN SAFETY AND QUALITY

How does workplace safety relate to a good-quality product? And which one has the higher priority? Nowadays, it is very popular to declare and endorse a *safety-first* priority among others in several organizations. Is it in such cases really an exclusive interest of the organization's management not to have any work-related accidents or does it conceal a deeper thought? All of these are issues dealt with by experts in the field of industrial activity management to achieve the highest prosperity and provide permanently sustainable development in an organization.

Based on the fact that part of a good-quality product (goods or services) is safety, it is also possible to state that safety culture is part of quality culture in every area of management activities. This statement might stipulate that quality is the prioritized feature of a final product in comparison to its safety.

Final product quality is the highest goal of any business. The connection between specific areas of complex management systems makes use of the advantage of a common goal—economically effective production with a competitive product. Creating efficient methods to achieve this goal requires constant inspection of meeting quality parameters by integrating the safety audit with the quality audit within the existing management system in an organization.

Through its products, an organization has to offer quality that is represented by a product that will not only meet every legislative requirement (if there are any), but also those stated by the customer. One of the significant requirements of a customer is risk minimization of the final product as well as within its production or, in case of a service, when it is provided. However, offering a good-quality product on the market does not mean only its compliance with parameters typical for its customer, but also that it will be delivered on time for an acceptable price (Figure 3.1). It is in fact

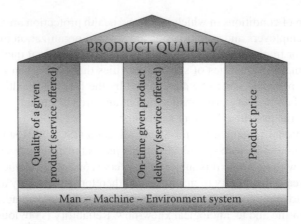

FIGURE 3.1 Product quality attributes.

the agreed delivery date that is in many cases the decisive criterion for the customer. On the other hand, in some cases it means that the producer is under pressure to provide flawless production, and no situations that might cause an increase in the risk of a negative occurrence in the form of an injury or machine malfunction. The reason might be the overload of two important elements in the Man–Machine–Environment system, which means an injury for the man or malfunction for the machine.

The current trend in the field of product design aims at a purposeful systematic approach. Issues of quality, safety, and reliability of the final product are the decisive entry parameters of product development. A satisfied customer is one of the main final parameters of product utilization, whereas their satisfaction means that the producer must create a good-quality, safe, and reliable product. Safe construction in this matter is above all systematic and methodical work.

The consequence of a work-related injury is the creation of workplace conditions that negatively influence production by failing to meet its required features on the outcome of the manufacturing process. The second parameter of quality is threatened due to production interruption, since it is not always possible to meet product delivery deadlines because of the documentation of a negative event (accident), the elimination of its consequences, and replacing the qualified labour force that suffered the accident. These aspects of manufacturing the final product only prove that its quality requirements are directly linked to requirements on risk minimization, and hence its safety. From a complex approach, it means a balanced consideration of quality culture principles within management activities with principles of safety culture. This is only one of the decisive procedures within integrated management systems or within methods of generic management that are nowadays increasingly being used. Underrating or neglecting the application of principles of workplace safety culture in an organization is in conflict with the effort to execute effective activities within quality management systems while observing principles of quality culture.

The outcome of handling attributes of safety culture and quality culture in an organization within its management means being aware of the following facts:

- Occupational health and safety, safety of technical systems, and manufacturing quality must be a part of the organization's development strategy as a tool to increase its prosperity.
- Prevention must have priority (from a complex point of view), and be implemented in every management activity of the organization.
- Responsibilities within occupational health and safety, safety of technical systems, and manufacturing quality must be clearly defined.
- Every employee in the organization must be concerned with occupational health and safety, safety of technical systems, and manufacturing quality.
- Principles of constant improvement as part of risk and quality management systems must be applied.
- The notion of *disaccord* must be perceived in the same sense in the field of quality and safety (although in safety, it might have a different name, e.g. injury, harm, damage, etc.).

Within its administrative activities, the company's management has a direct responsibility for activities within occupational health and safety, for safe operation and maintenance, and for the safety of manufactured products in compliance with respective legislative provisions. On the other hand, administration follows nonbinding requirements of standard in the case of quality management systems.

Management systems of occupational health and safety, safety of technical systems, or the increasingly widely used risk management system and quality management system are all part of integrated management systems in an organization, which result from the strategic goal of the organization, that is, achieving long-term prosperity. Implementing effective attributes of safety of technical systems, occupational health and safety, and manufacturing quality into management systems of a business presuppose the creation of organization culture through the integration of safety and quality culture, which are basic elements of the attributes of quality of life. Quality of life must result from constant training and education, especially in applying various methods of risk management systems and being aware of the fact that zero risk does not exist.

It is not only the manager's activities that include safety and quality culture in a successful organization, but also those of every employee. Safety and quality culture should be an integral part of lifelong learning (LLL) for both employees and employers. Safety and quality culture have to become part of every decision, not only within the professional activity of managers in successful organizations, but also in any area of social life in every community within a globalized world.

3.2 SAFETY, QUALITY, AND RELIABILITY FEATURES OF COMPETITIVE PRODUCTS AND MANUFACTURING PROCESSES IN GLOBAL MARKETS

New technologies and machine constructions might be described by a high level of complexity, which is constantly increasing. Their effect on the environment, ergonomic requirements, and technical solutions to eliminate failure of the human factor are all observed. Development trends in engineering expect the utilization of new technologies, new machine constructions with high-performance systems, and new materials.

Reliability is obviously one of the features of a safe product. It is the fulfillment of its working/user properties with respect to its stated requirements as defined in its technical conditions or manuals. It is, therefore, only natural that there are a number of common aspects and mutual links between quality management systems, safety of technical systems management, and occupational health and safety management.

Final product quality is the highest goal of any business. The connection between specific areas of complex management systems makes use of the advantage of a common goal, the economically effective production of a competitive product. Creating efficient methods to achieve this goal requires a constant check of meeting quality parameters by integrating a safety and environmental audit into the complex auditing system for the respective management system.

The current trend in the field of product design aims at a purposeful systematic approach. Issues of quality, safety, and reliability of the final product are the decisive entry parameters of product development. A satisfied customer is one of the main final parameters of product utilization, whereas their satisfaction means that the producer must create a good-quality, safe, and reliable product. Safe construction in this matter is above all a systematic and methodical work.

Safety, quality, and reliability of a product and its manufacturing systems require from every engineer participating in the project a summary of specific activities, from the product's feature definition to the execution phase of the manufacturing process, that happens at the same time in order to reach the best possible result—a good-quality and saleable product on the market. These activities include defining the goals and kind of manufacturing process, its planning, results, and test analysis, selection and qualification of components and materials, management of configurations, qualifications, supervision of manufacturing processes, and stating the procedures for implementing elements of quality and safety management. Some of these activities are difficult engineering tasks, while some are more tasks of coordination and supervision of the manufacturing process.

Nowadays, a mechanical device must be designed so that its operation and maintenance in expected technical conditions will not cause any hazard to elements of the Man–Machine–Environment system. The goal of approved measures must be the elimination or minimization of risk of any accident or material damage during the expected lifespan of a mechanical device, including its mounting or dismantling, even in operational conditions that do not comply with technical conditions for operation. It is necessary to monitor information describing operational conditions during the operation of a technical device in order to set values for its real technical condition, that is, to utilize methods of technical diagnostics collaboratively with the newest methods of experimental measurement.

Based on the aforementioned reasons, it is necessary to apply every efficient maintenance activity so that conditions for a technical device's (or even a complex technical device's) safe and reliable operation are provided during its lifespan, that is, to be able to constantly apply measures of technical and human risk minimization. The maintenance task has been modified due to the increasing importance of machine safety and maintenance. Machine malfunction occurrence might cause not only downtime, but also naturally the creation of specific types of hazards.

After every small, medium, or general repair, it is necessary to execute a risk analysis and to state whether the device is safe and risks during a following operation are minimized. Only safe and functional machines or complex systems might provide a complete circle of quality so that the result is a product that will be favourably utilized by a potential customer. Therefore, the employer, constructor and machine user, safety engineer, and the expert for quality management must collectively or separately identify, analyze, and quantify risks as well as execute measures for their minimization or elimination, that is, carry out all the activities within risk management.

In some concrete technological units, it is held that the more a machine or device is prone to failure, the larger the conflict between safety and quality is. This only naturally stipulates that *weak points* in complex or partial systems must be eliminated to increase their reliability. Increasing reliability therefore means the elimination of weak points and hence risk minimization, which leads to a product of higher quality that has a profitable financial effect for its producer.

In the past, sufficient agreement between defined (required) features of a product and the resulting safety and reliability parameters for a product or manufacturing process was deemed as a high level of quality. The increase in complexity of devices and systems, as well as a substantial rise in prices related to false operation in recent years, has ignited the creation of scientific branches that deal with issues of safety, quality, and reliability in terms of maintenance technologies. These days, it is expected that complex technical devices and systems, as well as manufacturing technologies, have the smallest number of systematic failures possible, and not only in time $t = 0$ (i.e., at the time of activation), but also in any period of the product's lifespan. Safety and equality enable optimization between the resulting parameters of product reliability and price during the product's *life cycle*. The parameters mentioned might be positively influenced as early as at the stage of creating the final product, for instance, by utilizing methods of concurrent engineering.

3.3 POSSIBILITIES OF INTEGRATING REQUIREMENTS OF SAFETY AND QUALITY IN THE CONSTRUCTION PROCESS

New technologies and machine constructions might be described by a high level of complexity, which is constantly increasing. Their effects on the environment, ergonomic requirements, and technical solutions to eliminate failure as a result of human factors are all observed. Development trends in engineering expect the utilization of new technologies, new machine constructions with high-performance systems, and new materials. One of the decisive criteria for their utilization is their safety in the Man–Machine–Environment system. It is important to create conditions for safe operation in the case of new machines and technological units as early as the development stage and then during construction. Furthermore, it is important to define the hazard rate in evaluating the safety of a technical system and machine construction as a whole. In order to assess the hazard rate, it is necessary to state the probability of its occurrence and assess the extent of possible consequences, that is, risk assessment. It is necessary to apply methods of quality and risk management to increase the quality, safety, reliability, and lifespan of machine construction. The level of these methods is influenced above all by the level of science and technology. Appropriate utilization of systematic engineering technologies is one of the ways to minimize risk and at the same time create conditions for higher quality in final construction production. It is a sophisticated systematic approach in the constructional stage of product design.

3.3.1 Quality in the Construction Process

Safety, and thus the quality of the product and its complex mechanical systems, require from every engineer participating in the project a summary of specific activities that happen concurrently, from the product's feature definition to the execution phase of the project, in order to reach the best possible result—a good-quality and saleable product on the market. These activities include defining the goals of manufacturing processes, development, results and tests analyses, selection and qualification of components and materials, management of configurations, qualifications, and supervision of manufacturing processes. Some of these activities are engineering tasks at the stage of development and construction, while others are tasks of coordination and supervision of the manufacturing process (see Figure 3.2).

In the 1960s, sufficient accordance between predetermined (required) features of a product and real resulting parameters was deemed as the high level of quality. The increase in complexity of devices and systems, as well as a substantial rise in prices related to errors as consequences of improper manufacturing processes, promoted the creation of scientific branches that deal with the issues of integrating elements of quality management, and hence safety, reliability, and maintenance, into the first stages of a final product's lifespan, that is, into its development and construction. The task of modern procedures in the construction stage is to ensure that machines and complex technical devices have the smallest number of systematic failures possible, and not only in time $t = 0$ (i.e., at the time of activation), but in any period of the product's lifespan as well. Application of elements of quality management in the machine's construction stage means reaching optimization between resulting requirements, that is, between safety, reliability, manufacturing qualities, and price during its lifespan. It is possible to achieve the aforementioned requirements by executing integrated systems, that is, integrating methods of quality management, environment management, and safety management into the system of concurrent engineering.

We might summarize the basic rules of applying elements of quality management in the construction stage of machines and technical devices in the following points:

1. The level of quality, and hence safety, has to be as high as is necessary to satisfy the customer's real requirements: *Apply the rule 'as high-grade as necessary.'*
2. Activities related to quality evaluation should be executed continually during each stage of a machine's lifespan, though especially at the stage of development and construction: *Retain the project manager before the prototype production is finished.*

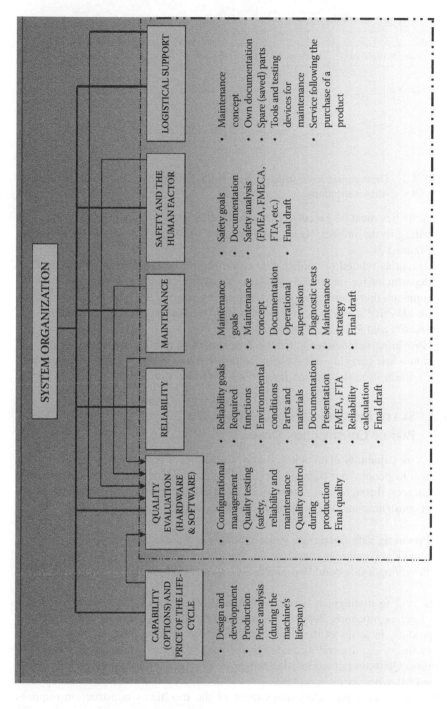

FIGURE 3.2   Elements of quality integration during the lifespan of a technical object.

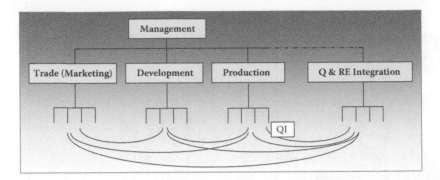

FIGURE 3.3   The organizational structure for quality evaluation (solid lines illustrate cooperation; QI = quality inspection).

3. Activities must be executed through the teamwork of all engineers participating in the project: *Apply the approach of concurrent engineering* (see Figure 3.3).
4. Activities related to quality evaluation should be supervised by a central department for integrated management systems (quality, safety, and environment) that coordinates activities in every stage of the project: *Implement an effective department for quality and reliability evaluation (Q&RE— Quality and Reliability Department).*
5. Providing independence for the department of quality and safety evaluation when carrying out reports for upper management: *Assign necessary capabilities to the central department of Q&RE* (see Figure 3.3).

3.3.2   SAFETY, RISKS, AND ACCEPTABLE RISKS AS A PART OF CONSTRUCTION DESIGN

Safety is the capability of an object, that is, a machine or any of part of it, not to cause any harm to people, nor damage (destruction) of material, nor other unacceptable consequences throughout its utilization. Integration of safety requirements at the stage of constructional design must be based on these two aspects:

• Providing safe operation, if the machine is operated in compliance with technical conditions stated by the constructor in advance
• Providing a safe condition, if a failure occurs in the machine

It is vital to distinguish between technical safety and reliability. A machine's safety level significantly depends on the required responsibility of a producer for their product, in terms of the European Union and specific national legislation. In most of the cases, drafted measures to increase machine safety also influence its reliability. Quantitative safety assessment is executed by estimating or calculating the risk, whereas its acceptable level is defined by the value of its acceptability. Risk utilization for safety assessment of the machine's construction requires interdisciplinary procedures, and must be solved through cooperation between engineers, psychologists, often even politicians, in order to find common solutions that lead to a satisfied customer at the end of this chain. In this matter, the choice of a suitable relationship between the possibility of failure (or a negative event) and the resulting consequence is crucial. Furthermore, it is necessary to consider various reasons for negative events within the Man–Machine–Environment system, as well as their consequences (situation, time, parties involved, man, consequence duration, etc.). It is possible to use statistical methods for risk assessment and evaluation of the possibility for negative event occurrence.

3.4   INTEGRATED MANAGEMENT SYSTEM

ISO 8402 defines the notion of *quality* as a summary of an object's features regarding its capability to meet stated and expected requirements. In this sense, we might mention that safety is one of the elements of quality. Hence, it is only natural that there are a number of common aspects and mutual links between quality management systems, safety management of technical systems, and occupational health and safety, that is, risk management, that have recently expanded in the area of environmentalism.

If these thoughts are appropriate, it is possible to formally integrate elements of environmental system management into quality management systems. Consequently, even elements of occupational health and safety management in the form of risk management (RM) systems according to Occupational Health and Safety Advisory Services (OSHA) 18001 or the MOP Directive might be included in quality management systems. Currently, quality management (QM) systems, such as ISO 9001, as well as environmental management systems (EMSs), for example, ISO 14001, are defined by individual legislative measures. An integrated standard that would define the three management systems—quality, environment, and safety—as a joint base for features of the final product in its various forms is not yet available. A graphic presentation of an integrated management system can be seen in Figure 3.4 and is also mentioned Chapter 1.

Integration of QM, RM, and EMS systems as a complex system has the following advantages:

- Provides a high level of quality as a basic factor of competitiveness in the market
- Increases the safety of technical devices and occupational health and safety with an aim to decrease the number of injuries and failures as a part of the strategy to guarantee product safety and the advantage of insuring strategies of a business
- Improves social acceptance of a business
- Minimizes the consequences of covering costs caused by failing to meet legal regulations in the field of occupational health and safety and environment protection
- Creates conditions to eliminate conflicts between companies and state inspection offices

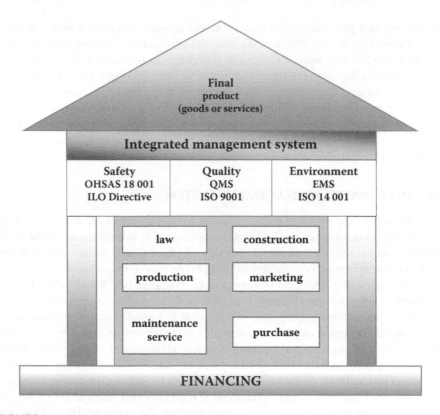

FIGURE 3.4   Scheme of the integrated management system.

3.4.1 MANAGEMENT SYSTEM INTEGRATION: EXAMPLE

We might demonstrate an example of mutual integration of activities from the fields of QM, RM, and EMS on piping systems, which can be seen in Table 3.1.

Transportation of media in the piping must be provided so that the properties of the medium in the outlet of the piping meet requirements that are defined in technical conditions for its utilization, which is within the field of quality management.

Features of the transported medium must be observed when drafting the piping, even in the case of a leak into the environment, which is the environmental aspect. Conditions for piping dimensions, as well as the choice of the production material, are defined based on an analysis of the aforementioned features. This stage is marked by the integration of technical safety and environmental aspects regarding the choice of piping material.

From the point of view of safety, operation of the piping also requires a definition of technologies for its surface customization and an appropriate choice of materials for its execution. This is again the integration of safety and environmental requirements.

Mounting the piping and technologies of its surface customization, as well as executing consequent maintenance activities, requires, once again, the application of activities within risk and environmental management.

TABLE 3.1

The Mutual Relationship of Tasks within an Integrated Management System

Function	Tasks in the Field of Quality Management	Tasks in the Field of Risk Management	Tasks in the Field of Environmental Management
Research and development of a device	Research and development of new concepts of quality management with the aim to define quality parameters	Research and development of new risk assessment methods	Research and development of environmentally appropriate goods and technologies
Design, planning	Conditions to meet product quality requirements	Find out what kind of product safety is required by the customer	State customer requirements on observing environmental aspects during production and recycling or destruction of a product
Manufacturing tools and maintenance	Maintenance aimed at providing quality and long lifespan of a technical device	Identifying and eliminating hazards due to malfunction of a technical device; integration of new elements of safety devices.	Utilizing tools and materials that meet environmental requirements; increasing the operating capability of a technical device based on its possible reconstruction; preventive change of loose or technologically dated parts
Material flows	Executing every activity within material flows to avoid devaluation of transported objects	Providing safety in material transportation, identifying hazards, and execution of measures to eliminate them	Choice of short transportation distances, avoiding unnecessary transportation, utilization of transportation devices that observe effects on the environment during their operation

An example of implementing each activity within an integrated management system related to some of the company's functions is presented in Table 3.1.

3.5   OCCUPATIONAL SAFETY AUDIT

In the area of business economics, the notion of an *audit* means verifying the condition of a business (or company) or an integral part of one of the organizational units. Results of such verification must be recorded in various forms.

An occupational safety audit or safety audit comprises the experiences of managers and experts in the field of occupational health and safety during the examination of a business, inspection and control of workplaces and technical devices, and during training and seminars.

The following are the goals to be achieved in a safety audit:

- Danger and hazard identification
- Creation of conditions for risk management
- Drafting a more effective method for occupational health and safety activities
- A positive influence from other levels of a company's management in order to increase the quality of the final product as a part of the integrated management system
- Achieving lower costs for the business

A safety audit is one of the basic elements of safety culture. Safety culture presupposes the creation of working conditions in which occupational health and safety is understood as a joint task of employers and employees on every level of the company's management.

3.5.1 BASIC ACTIVITIES WITHIN SAFETY AUDIT

An essential element of a safety audit is the examination of a business or its individual parts. It is necessary to broaden the examinations to a complete safety audit if the knowledge gained during these examinations is insufficient, that is, information is not detailed and complete. The following points sum up the core of examination activities, which compare the actual conditions with the required conditions subject to laws, provisions, and directives:

- Assess imperfections in occupational health and safety by identifying dangers and hazards, and develop a consequent risk assessment.
- Carry out measures for perfecting activities in risk management.
- State clearly how these measures will be executed.

Auditors and audit groups investigate whether a business or any of its organizational parts meets the requirements of occupational health and safety, and in case of technical devices, the reliability criteria as well.

The company's management might make the business 'safer,' minimize risks, and hence be economically more successful, but only if they become familiar with such tasks and convince their employees about the importance of the success of every activity in occupational health and safety.

Audit execution is conditioned by the reason and goal that is set forth to be reached. Auditing activities are carried out as follows:

- They are announced to operators in advance or unannounced.
- They are done regularly according to set deadlines or irregularly.
- They are systematic or random.

They are based on a concrete reason in these instances:

- Mounting a new device
- Putting the device into operation
- Reconstruction, large repairs, or maintenance
- An increase in the number of accidents or failures
- Legislation change
- A decision by the company's management to increase the level of occupational health and safety

It is possible to expect detection of real operational conditions during unannounced examinations. The audit's success is negatively influenced by the fact that a director does not have enough time for the auditor.

Examinations announced in advance might cause 'artificial' operational conditions. However, directors, as well as employees, will be directly included in the audit's activities and will not feel distrust.

A director's participation in the audit's activities increases the authority and importance of risk management in a business.

3.5.1.1   Auditor Appointment

Provided that a company's management decides to execute a safety audit, it is necessary to execute an important decision: *Who will be the auditor or who is suitable to execute auditing activities?*

Auditor (or audit group) appointment is conditioned by tasks, the extent, and the aims of the safety audit. The first basic auditing activities are executed by professionally qualified directors of the company itself. An increasing number of employees, as well as the extent and complexity of technologies within a business, require an increasingly higher need to have independent auditors or specialized auditing agencies to execute the safety audit. The company's management has to participate in the execution of the safety audit by experienced experts who are not dependent on it in any sense.

It is necessary to engage an auditing staff if the tasks of the safety audit are too complex and vague, and actions must be executed in a specified period of time. This staff includes experts from various areas, although there should not be more than five people working on it.

A safety audit is distinguished by:

1. Capability, which includes:
   - Specific knowledge
   - Success of a director
   - Experience in the field of occupational health and safety
2. Personality features:
   - Objectivity
   - Communicativeness
   - Ability to recognize a certain situation
   - Patience

3. Independence:
   - Equal partner
   - Freedom of decision making

3.5.1.2 Safety Audit Execution

It is appropriate if the company's management calls upon every director and other employees to actively cooperate, even before the auditing activities begin. A safety audit cannot be a concealed activity. Only an informed employee can comprehend the importance and advantages of the audit.

First, it is necessary to execute an examination of the business. Auditors gain knowledge about specific parts of the material flow, about technical devices, and the share of the human factor in individual parts of manufacturing technologies. Questionnaires prepared beforehand are the basic tool to lead specialized interviews. Their aim is to recognize strengths and weaknesses of the company. Auditors must observe these rules to fulfill tasks that result from applying questionnaires:

Choose a period of intense working time when questionnaires are to be filled in:

- High degree of working time utilization
- Shift changes
- Night shifts
- Removing failure consequences
- Maintenance

Accumulate real information:

- Respond to every notice related to issues of occupational health and safety
- Confirm information by facts
- Monitor working procedures
- Attempt to objectify information
- Prevent criticism
- Have time to lead specialized discussions
- Consider employee behaviour

Questionnaires could contain general questions, such as:

1. Are risk minimization measures adequate and purposeful?
2. Do the questionnaires obstruct employees at the workplace?
3. Is a smooth material flow provided within the company's logistics?
4. Are the responsibilities and the conditions of their mutual coordination within the management system of the business clearly stated?
5. Are employee suggestions to improve occupational health and safety conditions observed?
6. Do the employees have sufficient knowledge and experience?
7. Are some of the employees overloaded or, on the other hand, idle?
8. Are some of the employees distinguished by 'hazardous' habits?

9. Do the superordinates affect 'hazardous' activities?
10. Are the employees trained on how to behave in case of a breakdown or failure?
11. Are internal relationships functional?

The safety audit helps to recognize weaknesses in a business. An audit does not serve to find who is guilty. All interviews must be done objectively and their content cannot blame anyone.

3.5.1.3   Safety Audit Evaluation

An auditing questionnaire must take into account the company's actual condition. This condition is then compared to one that is in compliance with legislative enactments in a given society at a certain period of time.

It is necessary to exactly formulate the differences and then draft measures for their minimization or elimination. These conclusions must be dealt with by responsible directors. If the company's management associates themselves with the auditor's conclusions and recommendations, that is the most favourable situation.

Results of the safety audit can be summarized into points that are included in the activities within risk management.

A check of executed safety audit provisions is carried out during the evaluation of results. A plan to execute recommendations is carried out after the audit's conclusions are released. If the recommendation cannot be executed or it is amended, it is necessary to provide the reasons that led to such a situation.

3.5.1.4   Safety Audit Types

According to the type and area of execution, safety audits are classified as follows:

a. Engineering audit (e.g. machines and machinery, facilities)
b. Manufacturing process audit (e.g. lighting, fire, explosion, technologies of hazardous substances, and transportation of heavy loads)
c. Management audit (e.g. personal and collective protection, maintenance, first aid, employee participation, respective legislation accordance)

An engineering audit, as well as a management audit, might be the result of a breakdown or failure.

3.5.1.4.1   Engineering Audit

The first stage of this audit is usually aimed at technical devices, since various types of hazards occur due to their operation, whereas the potential of these hazards is defined by engineering risks. The following might be objects of an engineering audit:

• Working facilities (e.g. transportation machines, cranes)
• Mechanical systems (e.g. packaging machines)
• Technological parts of a business (e.g. filling line)
• Complex plants (e.g. paint shop)

An engineering audit is executed by specialized engineers, directors, and safety technicians, usually as members of an audit group or individually. The following areas are recommended for engineering audit execution:

- Industrial machines
- Electrical machines and devices
- Fire protection devices
- Scaffoldings
- Lifting and manipulation devices
- High-pressure cleaning systems
- Personal protective aids
- Transportation systems
- Workshops

It is necessary to execute an engineering audit at every stage of the examined object's lifespan. The following might be included in the questionnaires:

- Working movements
- Noise and vibrations
- Temperature and air conditioning
- Electrical equipment and gear
- Hazardous materials
- Hazardous activities of facilities
- Accordance with respective legislation, standards, directives, and regulations
- Functionality of safety and protection facilities within a technological unit
- Statistical evaluation of technical reasons for specific accidents and breakdowns

3.5.1.4.2 Manufacturing Process Audit

The aim of a manufacturing process audit is above all to recognize the consequences of changes or negative events on the manufacturing process. The audit is executed by teams of experts from areas of production planning, machine safety, and occupational health and safety, as well as directors of a particular area of production. The audit is executed, for instance, in the following areas:

- Breaking in manufacturing technologies
- User manuals
- Waste disposal systems
- Measuring systems for dangerous substances
- Communication systems
- Software safety

It is not only the manufacturing process that is checked during an audit but also the areas directly related to the manufacturing process, for example:

- Employee degree of education
- Processes of cleaning and servicing machines within the complex technology
- Maintenance and failure elimination

Methods for preparation, progress, and evaluation of the manufacturing process audit are identical to other forms and types of audits.

3.5.1.4.3   Production Management Audit: Safety Audit Aimed at Risks in Management Systems

Manners of managing a business by its superordinates, risk management organization (management system), and effectiveness of occupational health and safety activities are analyzed within the production management audit. Execution of this audit usually happens either regularly (e.g. once every five years) or due to unexpected circumstances (e.g. a sudden increase of accidents at one workplace).

A production management audit might help to gain information about:

- Safety awareness of directors
- Concepts and strategies within risk management
- Condition, extent, and quality of foundations and documentation for successful occupational health and safety management

A production management audit usually includes the following areas:

- Use of safety regulations
- Verification of employment contracts
- Activities in occupational medicine
- Education
- Corporate breakdown plans
- Execution and documentation of safety examinations and checks
- Employing workers of other companies
- Utilization of personal protective aids
- Information flows and communication forms
- Maintenance and repairs programs
- Order and cleanliness
- Occupational health and safety philosophy and its aims
- Ways of investigating occurrence, progress, and consequences of accidents and breakdowns
- Manipulation of dangerous substances
- Safety in transportation and manipulation of material

It is important to include the following as well:

- Fire protection concept
- Waste disposal programs, including their separations (environmental aspect)

- Programs to minimize ergonomic impacts (e.g. noise)
- Environment protection concept

A production management audit analyses in individual plants of the company are executed by respective directors (e.g. operating manager, foreman), occupational health and safety experts, as well as by experts in the area of the technologies investigated. An employee representative might be also a member of the team, if necessary.

A production management audit is significantly important for risk management, since it:

- Clarifies readiness and it will be up to the company's management to support the increase of occupational safety culture
- Documents the readiness of the company's management to conform their current attitude toward issues of occupational health and safety to criticism in terms of an example's function
- Contributes to an ever-intensive engagement of upper managers in issues of risk management

Independent, highly qualified auditors must be charged with the task of executing analyses within the production management audit.

BIBLIOGRAPHY

Pačaiová, H., Sinay, J., and Glatz, J. *Bezpečnosť a riziká technických systémov* SjF TUKE Košice Edition, Vienala Košice 2009, ISBN 978-80-553-0180-8-60-30-10.
Hrubec, J. et al. 'Integrated management system—2007', in *Research and Development Projects*, Košice: HF TU, 2007, pp. 33–34, ISBN 9788080738303.
Sinay, J. 'Audit bezpečnosti práce ako súčasť komplexného auditu prevádzok', Conference: Current issues of occupational safety, XI. International Conference, VVUBP Bratislava, 1998, pp. 50–59.
Sinay, J. 'Bezpečný podnik: Moderný systém integrovaného riadenia podniku', Specialized seminar, OHS as a Central Part of Integrated Management System," VSŽ, VaPC Košice, April 2000, pp. 60–77.
Sinay, J., Markulík, Š., Pačaiová, H. 'Kultúra kvality a kultúra bezpečnosti: Podobnosti a rozdielnosti', in *Kvalita Quality 2011*: 20, International Conference: 17, 18.5.2011, Ostrava, Ostrava: DTO CZ, 2011, pp. A21–A24, ISBN 978-80-02-02300-7.
Sinay, J., Oravec, M., Kopas, M. 'Manažment rizika ako súčasť integrovaného systému riadenia kvality', 3. International scientific workshop, Machine quality and reliability, in *International Engineering Expo 98*, Nitra, May 1998, pp. 92–95, ISBN 80-7137 487 3.
Sinay, J., and Pačaiová, H. 'Risikoorientierte Instandhaltung', in *Technische Überwachung* 44,č. 9, 2003, pp. 41–43, ISSN 1434-9728.
Sinay, J., Pačaiová, H., and Kopas, M. 'Risk management-Component of Total Quality Management/TQM/', Konf. HAAMAHA 98, Hong Kong, July 1998, pp. 316–320, 30% in *Science Report—Project PL-1, Metrology in Quality Assurance Systems*, CEEPUS Program, Kielce University of Technology, Poland, 1998, ISBN 83-905132-9-3.

Sinay, J., Pačaiova, H., and Oravec, M. 'Posudzovanie rizík, základný parameter konkuren-cieschopnosti podnikov', *Acta Mechanica Slovaca* 12, April 2008, pp. 51–56, ISSN 1335-2393.

Winzer, P., and Sinay, J. *From Integrated Management Systems towards Generic Management Systems: Approaches from Slovakia and Germany,* Shaker Verlag, Aaachen/BRD, 2009, ISBN 978-3-8322-8508-1.

4 Theory and Selected Applications of Risk Management

The process of globalization of industrial activities and global labour markets has significantly influenced the field of occupational health and safety, including risk management.

The complexity of working and production activities of entrepreneurial entities nowadays calls for the introduction of the systematic organization of work and of control mechanisms, which will ensure correct functioning of all the production and managerial processes. The quality of company management is a presupposition for fulfilling production goals, but it is also a criterion of competitiveness, a requirement for success in the market and a token of the reliability of a business partner. It has become a natural thing for a business partner to examine, by means of customer's audit, the quality of the organization of their contractors' work. This kind of examination includes checking the quality level of the production process control, financial process control, quality and environment control, and occupational health and safety management. If a company takes serious care about its future, it strives to transparently implement the control systems in respective areas.

Instructions on how to implement the risk control system are available, for example, in the most widespread document on the subject, Occupational Health and Safety Advisory Services (OHSAS) 18001 (and its interpretation, OHSAS 18002), Chapter 2. It is not a regular International Organization for Standardization (ISO) standard, as is commonplace in the field of quality and environment controls, but a majority of certification authorities worldwide follow the standard.

Mobility of international capital and foreign managerial structures has had a positive influence on a broader introduction of the occupational health and safety control systems. There are ever more frequent requirements within the supplier–customer relationship to present evidence of the safety control systems' efficiency.

The implementation of an appropriate occupational health and safety (OHS) management system by employers has its purpose in practice. By creating a suitable mechanism that will assure the proper functioning of an entrepreneurial entity in the field of OHS, the quality of OHS may constantly be improving. All of this is of high economic importance, since dealing with the issues of health and safety, and in a broader sense, creating favourable working conditions and relations, result in the optimization of working processes at companies and have a positive economic effect. This also helps cost cutting, and improves work productivity, efficiency, and quality, which brings prosperity to a company and, eventually, to the whole society.

In addition, it includes an important human aspect that contributes to cultural and social standards of both the company and state.

4.1 BASIC PROCEDURES IN RISK MANAGEMENT

Principally, basic procedures within risk management, including occupational health and safety management, may be divided into the following steps:

Step 1: Preliminary status analysis and, most of all, the possibility of proper division of the defined system's structures. The following must be taken into account when dividing the system into structures:

1. Analyzing the material flow within production technologies (the inputs and outputs of the main production process)
2. Defining the staff migration during all operating modes, including the machinery setup and maintenance
3. Following the legal enactments (in compliance with ISO 9001, ISO 14001 with emphasis on the provisions dealing with safety and illegal operations)
4. Determining technologically problematic conditions and nodes at the plants analyzed; defining potential dangers and ensuing hazards

Step 2: Filling out the catalogue sheet for the profession or activity analyzed, depending on the kind of operation and legislation. This is to provide an objective assessment of the actual conditions. It is not advisable to create a universal catalogue sheet since each plant is unique with its specific parameters. In the process of creating the catalogue sheet, it is necessary to consider whether it is to be created for a profession or a working activity.

It is important that there is a possibility to identify dangers and associated hazards based on the content of the catalogue sheet, and that there is valid legislation assigned to those.

Step 3: The application of a suitable method for the risk estimation or assessment based on the potential hazards described in the hazards catalogue (a respective catalogue sheet, a specific profession, a specific group of hazards).

A particular risk that exists within the operational process and is a function of the individual system components is assigned quantitative criteria, for example, numeric values. These values further contribute to the determination of the extent of the resultant risk.

Step 4: The process of suggesting measures to reduce the existing risks. The form of recording the measures is normally defined by individual companies' internal regulations. When summarizing individual risks for respective professions, it is possible to define the order of importance for the measures within the Man–Machine–Environment system.

Based on the data collected in the process of risk assessment, as well as on the risk values in individual professions, it is possible to propose specific measures for a given profession or a particular workplace and, moreover, to propose global measures related to several professions and workplaces.

Some other factors must also be considered when assessing risk. Those include, in particular, uniformity in the interpretation of terms that define individual parameters, and definitions of the procedures for risk elimination.

4.2 CAUSAL RELATION OF FAILURE OR ACCIDENT OCCURRENCE: WILL A FAILURE OR ACCIDENT OCCUR BY COINCIDENCE?

A failure or accident is a sudden, undesirable, and unexpected event that could harm people and damage technical systems, and this could result in the interruption or breakdown of the planned operating conditions. In order to prevent and eliminate faults and accidents, the systematic analysis, identification, and description of the causes and progress of such events are necessary. The information collected provides for the proposal of measures to prevent the occurrence of the events. Unless identified, reduced, or eliminated on time, the causes of faults and accidents that have occurred become the causes of potential negative events in the future.

Figure 4.1 shows the causal relation of failure or accident occurrence, which includes the following five time-dependent stages:

1. Danger
2. Hazard
3. Initiation
4. Damage
5. Harm

The diagram is based on the fact that the functional relation formed this way applies to all kinds of faults and accidents; therefore, their occurrence is not coincidental, but based on certain regularities. In practice, it is crucial to get familiar with the progress of this causal relation in order to create a system for interrupting the relation, and thus preventing the failure or accident occurrence.

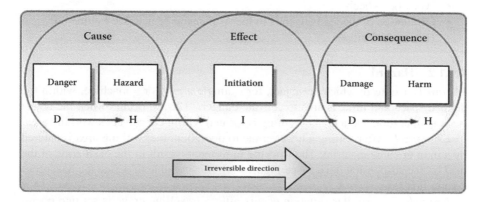

FIGURE 4.1   Causal relation of failure or accident occurrence.

The mutual, causal relationship of a negative event occurrence and its individual stages is shown in Figure 4.1; the process is irreversible, that is, it happens in one direction only. The individual stages of the process occur in sequence.

4.2.1 DEFINITION OF STAGES IN CAUSAL RELATION OF A FAILURE OR ACCIDENT OCCURRENCE

Individual mutual relations within the causal relation of failure or accident occurrence will be herein demonstrated on a lifting machine as one of the most frequently used pieces of machinery within various types of logistic systems. It is a machine that is included in the group of technical equipment with high risk levels involving the gravitation effect while the machine is operating.

4.2.1.1 Danger

Danger is the quality of a machine, subject, technology, event, or man (in case of civil security) to cause damage followed by harm—a negative event.

If danger is not activated, that is, put into operation, it is of no interest as far as scientific analyses are concerned. Under these circumstances, there will be no damage or harm to the given object.

Example 1: A tower crane. Its structure has features that may cause a collapse, which leads to human and technical damage. However, unless the crane is used to hoist heavy loads, with no strong winds and no force majeure (e.g. an earthquake), damage and harm will not occur (see Figure 4.2).

Example 2: The dynamic properties of a travelling crane as a flexible structure cause oscillations, which are further transferred to the operator's area, the cabin. The oscillation occurs as a result of a non-stationary event during the travelling crane's operation. The following are among the dangers:

* Inertial effects of working movements—starting, braking
* Stochastic geometric deviations of the crane track—the railway unevenness
* Worn-out wheels
* Oscillation of the load attached to the suspension device (see Figure 4.3)
* Hitting the bumper

4.2.1.2 Hazard

Hazard is a state in which an object, for example a man or a machine, within the defined space and time, is able to activate dangers. Hazard occurs when the object is set to operation and a man or a thing is located within the subject's working area.

Example 1: After setting a tower crane to operation, the working area is entered by a man or some material. The working movements result in the oscillation of the load on the suspension device (see Figure 4.4). It is possible to define the hazardous area in this case.

Example 2: A similar situation occurs after a travelling crane is set into operation, where the working movements result in the oscillation of the load and, thus, in hazards to materials and people if they enter the working area (see Figure 4.5).

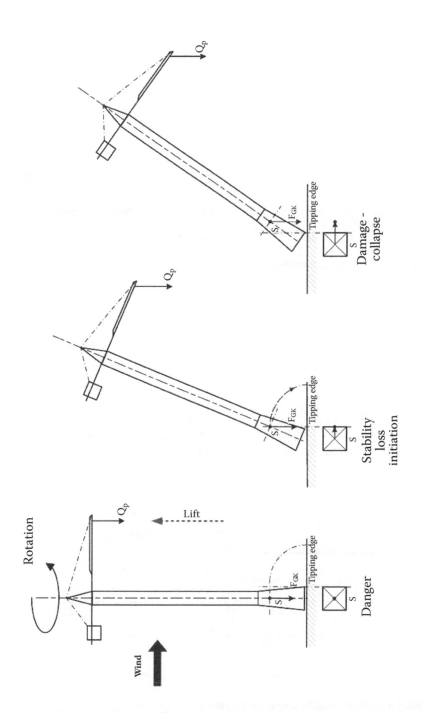

FIGURE 4.2 Scheme of lost stability of a tower crane.

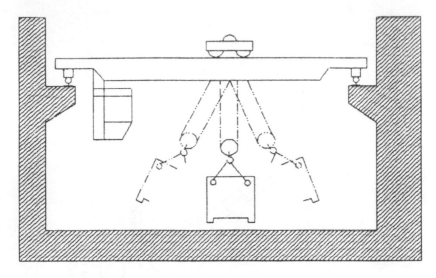

FIGURE 4.3   The load oscillation as the identified danger.

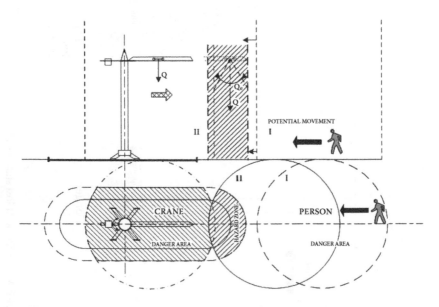

FIGURE 4.4   Hazard scheme during the operation of a tower crane.

FIGURE 4.5 Hazard scheme during the operation of a travelling crane.

Special literature and standards of some countries include the terms *dangerous event* and *dangerous situation*, which may as well be associated with hazards.

4.2.1.3 Risk

The term *risk* is used in all languages, with a relevant definition as a hazard degree (potential) and its importance is unequivocally defined as a relationship between the probability of a negative event occurrence—harm, injury, accident, P, and the consequences following the damage, injury, accident, C as in:

$$R = P \times C \tag{4.1}$$

Risk represents the degree of hazard, and both parameters are dependent on a number of conditional factors, which is reflected in the content of procedures for its quantitative assessment. Danger, hazard, and risk are associated with a single negative event and are directly interrelated (see Figure 4.6).

Provided that hazard is considered a basic term for a certain negative event, then danger forms a source of hazard, and risk determines the degree (potential) of the hazard.

4.2.1.4 Initiation of a Negative Event

This is a crucial stage in the causal relationship of a negative event occurrence. During this stage, an impulse for interrupting the system balance appears. The impulse may be caused by a man (e.g. by performing an inappropriate operation), technology (e.g. faulty safety devices, the inappropriate suspension system for the seat of a travelling crane operator, eventually caused by a man), and environment (e.g. undefined non-stationary events during a travelling crane operation, wind affecting the hoisting mechanism operation, seismic effects, or geological structure of the soil). In the

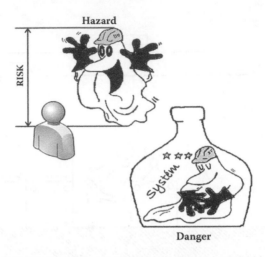

FIGURE 4.6   Relations among danger, hazard, and risk.

majority of cases, defining the beginning of initiation is of a very random nature, and from the point of view of preventive measures, the initiation is used least effectively. The damage initiation, for example, in the case of oscillation effects on a man in the travelling crane cabin, is defined by the boundary values of oscillation frequency and actual acceleration. As a part of mathematical analyses, the time period, one of the dynamic system parameters, is defined as the period during which the operator is exposed to the cabin oscillation. The moment at which these values are exceeded could be considered as the moment of initiation of the negative event occurrence, and in this case, health problems were a result of the oscillation.

4.2.1.5 Damage

This is an active stage of the causal relationship in a negative event occurrence. As for the effective prevention, it is crucial at this stage to exactly define the progress of this process if possible, normally as the function of time or the period of utilization. The progress of damage processes could be described as fluent damage (stochastic or deterministic; see Figure 4.7a and Figure 4.7b) and of saltus nature (e.g. fatigue damage of iron material, damage as a result of wear, corrosion; see Figure 4.7c) (e.g. damage of fragile materials such as glass, forced failure of a steel rope as in rupture, electric circuit failure).

The process of damage as a stage at which preventive measures may be introduced (the interruption of the causal relation of a negative event occurrence) must be defined as a time function, since the selection of effective preventive measures is conditioned by the understanding of the function.

The following conditions must be fulfilled for effective preventive measures applicable at the damage stage:

- The occurrence, type, and progress of damage must be defined (e.g. the progress of fatigue fracture of a steel structure).

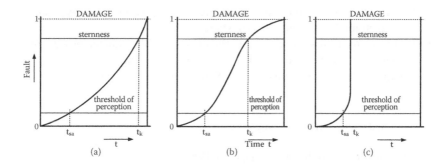

FIGURE 4.7 Mechanism of damage progress.

- The object assessed must be fully functional during the process of damage.
- The process of damage must be quantitatively assessable by means of measuring chains.

Within the risk control of industrial equipment, knowing the progress of the damage stage is crucial for selecting the strategies for maintenance, inspections, and checks, that is, the means to reduce the risk.

There are basically two kinds of progress of damage processes:

1. Progressive
2. Digressive

From the point of view of the efficiency and risk elimination preventive measures, the process in which the time for reaching the boundary for negative event perception t_{sa} and the time prior to the failure occurrence, that is, reaching the critical state t_k, are the longest possible, represents the progressive damage process (see Figure 4.8).

The following priority activities are necessary to be carried out in relation to effective preventive measures:

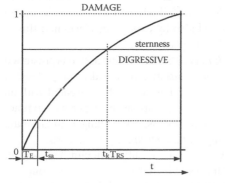
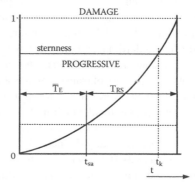

response time

FIGURE 4.8 Damage processes.

- Defining the damage processes as progressive, if possible, as early as at the design, projection, or construction stages
- Proposing measures for decreasing the intensity of the damage process (e.g. modern maintenance methods for industrial equipment)
- Creating redundant systems (e.g. airplanes, mechatronic systems)

4.2.1.6 Loss

Loss is defined as (also in other languages):

- a physical injury or health damage, and/or
- machine faults or breakdowns resulting in the loss of a subject's functional ability (e.g. the functional ability loss of machines and machinery complex systems or damage to health and environment).

4.2.2 NOTES ON TERMINOLOGY USAGE WITHIN THE CAUSAL RELATION OF FAILURE OR ACCIDENT OCCURRENCE ACCORDING TO VARIOUS STANDARDS

Standards are not and must not be binding. They represent merely a minimal standard that is followed by experts wishing to be 'on the safe side' when being short of other scientifically justified procedures.

Despite this, it is advisable in relation to possible discrepancies in the terminology definitions, to present readers with some notes on the meaning of selected terms based on the standards that are valid in the Slovak Republic, and in a majority of cases, in compliance with the European standards.

4.2.2.1 EN ISO 14121-1 Machine Safety Risk Assessment Principles

This is basically a Slovak translation of an English edition of the ISO 14 121-1 standard that has been elaborated by the technical committee of ISO/TC 199 Machine safety in cooperation with the technical committee of European Committee for Standards/Technical Committee (ECN/TC) 114 Machine safety, whose headquarters are located at DIN—a German standard-setting institution; and German is the language in which the European Standards (EN) ISO 14121-1 standard has been written. The standard has replaced the Slovak Technical Standard European Standards (STN EN) 1050 standard elaborated also by experts in Germany; this is an 'A' standard, i.e., an elementary (general) safety standard.

The standard treats the terms *hazard* and *danger* as equal (Article 3.2). It is presumed that in the process of translating the standard, the translation of Slovak *nebezpečenstvo* and *ohrozenie* were based on English *hazard*, which has not yet been treated with an unambiguous equivalent within the international community, as the experts from practice as well as scientists use this term for *ohrozenie*, that is, active danger. In German, both terms are clearly defined, *Gefahr* for danger and *Gefährdung* for hazard.

Danger (also Chapter 3.2.1) as the quality of a machine, subject, technology, event, and a man (civil security) to cause damage followed by harm—a negative event.

If danger is not activated, that is, it is not put into operation, it is of no interest as far as scientific analyses are concerned. Under these circumstances, there will be no damage or harm of the given object.

- A *dangerous area* is a hazardous space. In the German original, this term equals the term 'dangerous space'!
- A *dangerous event* is an event associated with a hazard, the activation of danger; it is an event that may cause harm!
- A *dangerous situation* is a state where one or more persons are exposed to at least a single, standard hazard.

4.2.2.2 EN ISO 12100-1 Machine Safety, Basic Terminology, General Principles of Machine Design—Part 1 Basic Terminology

Neither the principles nor the standard distinguish between *danger* and *hazard*. The other terms are identical to those of the EN ISO 14121-1 standard.

4.2.2.3 OHSAS 18001

This standard, the most frequently used one for occupational health and safety management, defines both *danger* and *hazard* as 'a potential source, situation or activity that may cause an injury, damage to health, or the combination of these.'

It is, therefore, obvious that the inconsistency and ambiguity of the basic terminology definitions within occupational health and safety management, as well as risk management, may lead to wrong conclusions related to the process of defining the sources of undesirable (negative) events within the analyses of damage and harm occurrence. The procedures for determining effective principles for reducing risk in a variety of industries and walks of life, including private life, as applied at the Department of Safety and Quality at Technical University's Faculty of Mechanical Engineering in Košice, are based on the usage of the unambiguous definitions stated in the analysis of causal relation of the occurrence of a fault or injury, a negative event, in general.

4.2.3 THEORY OF MEASURES TO PREVENT FAILURE OR ACCIDENT OCCURRENCE AS A BASIS FOR EFFECTIVE PREVENTIVE MEASURES

The fundamental purpose of all preventive measures within the effective system of risk management, as well as occupational health and safety management, is to first of all analyze all the stages of causal relation of a negative event occurrence during the technological lifespan of machines and machinery systems in all areas of their technological lifespan, their curriculum vitae, technologies, materials, working environment, and environment as such, in order to develop and design procedures that will allow for the interruption of the casual relation in its early stages. Each shift of the interruption point toward the harm, that is, the damage stage, means limited efficiency and increased costs (e.g. withdrawing cars to be repaired; Toyota in early 2010). This process is represented by the scheme in Figure 4.9.

The preventive measures against negative event occurrence may be based on the following (see Figure 4.10):

- Harm analyses, with an emphasis on defining the hazard or dangers, which is in a certain case a more effective method, but it is based on a negative event already occurring (marked as 1 in Figure 4.10)

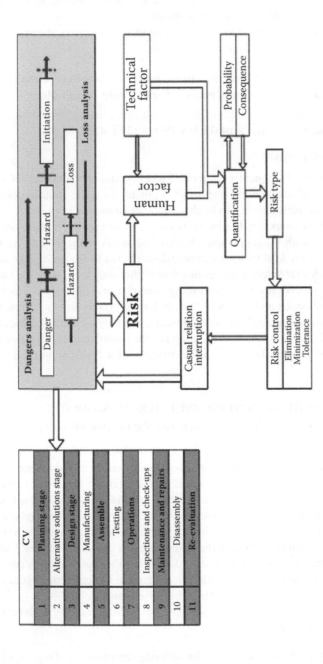

FIGURE 4.9 Scheme of risk reduction process. Reprinted with permission from Waldemar Karwowski and William S. Marras, eds., *The Occupational Ergonomics Handbook* (Boca Raton, FL: CRC Press, 1998), 1921.

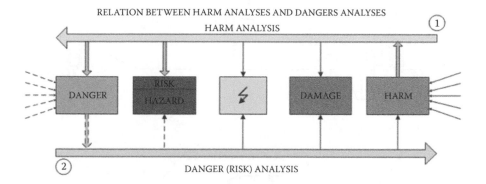

FIGURE 4.10 Possibilities for risk analysis.

- Hazards or dangers analyses, with an emphasis on the occurrence of potential harm, which is a method that does not assume the actual occurrence of an accident (marked as 2 in Figure 4.10)

Nowadays, with the existence of various methods to model and simulate machines, processes, and human behaviour using IT, including virtual reality (VR), there are a growing number of instances where analyses based on the early stages of causal relation are utilized.

4.2.4 Example of Controlling Causal Relation for the Risk: Crane Operator's Illness as a Result of Travelling Crane Oscillation

Regarding the interruption of the causal relation of an occupational disease for travelling crane operators, it is necessary to take into account the fact that the force flow is moving from its source along the whole construction of the machine as a result of working movements (see Figure 4.11). The working movements of the travelling crane result in the transfer of oscillation into the operator's area, the cabin. The mechanisms keeping the cabin attached to the travelling crane's steel body are typically fixed; this causes direct transfer of the oscillation to the cabin and, thus, onto all of the operator's body parts.

a. *Interrupting the causal relation at the damage stage* is plausible only if, following the occurrence of the initial signs of the negative effects of the oscillation on the lifting machine operator, the crane operator is prevented from performing his duties. It is important to ensure effective diagnostics of the operator's physical condition and determine sufficient intervals for regular health checkups.
b. *Using the initiation stage to interrupt the causal relation* is conditioned by the installation of an online system that monitors operating conditions and registers the actual working time and workload of the crane, as well as the oscillation frequency and its characteristics. Following the evaluation

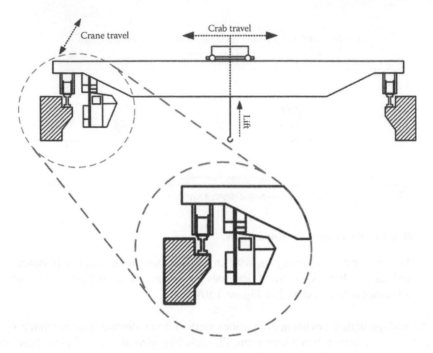

FIGURE 4.11 Travelling crane design with positioning of the cabin. Reprinted with permission from Waldemar Karwowski and Gavriel Salvendy, *Advances in Human Factors, Ergonomics, and Safety in Manufacturing and Service Industries* (Boca Raton, FL: CRC Press, 2011), 826.

of the parameters as a multicriteria function, the intervals for the crane operational duties are determined in order to avoid carrying out the crane operational activities for a period of time that exceeds permissible values.

c. *Interrupting the causal relation at the hazard stage* is an ideal procedure to be used at the machine design stage. This group of measures includes:

- Active and flexible cabin suspension: reducing the oscillation effects on operators
- Application of a remote control for the crane: excluding operators from the force-flow system and keeping them away from the body of the crane, but placing them within its working area
- Using automated travelling cranes: keeping operators away from the body of the crane and from its working area

These measures may be considered the most effective; economic parameters must be taken into account when applying the measures.

4.3   INTEGRATED APPROACH TO RISK MANAGEMENT

Current European legislation in the field of technical equipment safety is based on Directive 42/2006/EC on machinery standards harmonization. According to the directive, Annex 1, Subsection 1.1.2, Paragraph b, a producer will:

- Eliminate or reduce risks as far as possible (integration of safety into machinery design and construction)
- Take the necessary protective measures in relation to risks that cannot be eliminated
- Inform users of the residual risks due to any shortcomings of the protective measures adopted; indicate whether any particular training is required and specify any need to provide personal protective equipment

The measures applied must be focused on the elimination of risks during the expected lifespan of machinery, including the time of transport, assembling, and disassembling, and the time of shutdown, disposal, or recycling.

These requirements define the activities of a designer, manufacturer, and user of a machine, which include performing all the procedures within the risk control systems. This is based on the principle of application of the causal relation of accident or failure occurrence as a part of the risk control systems at all the stages of a machine's technological lifespan.

Based on the content of the enactments, it is obvious that, for the sake of the safe usage of a machine during its lifespan, high demands are placed upon the design stage of the machine as the first stage of a machine's technological lifespan and as the stage with the most favourable conditions for applying the measures to interrupt the causal relation of an accident or failure occurrence. In relation to these requirements, it is possible to define the effective measures for risk control for individual stages of the given equipment's technological lifespan.

Figure 4.12 demonstrates the relation between the design stage and the usage stage of equipment in relation to the extent of risk and the methods for risk assessment.

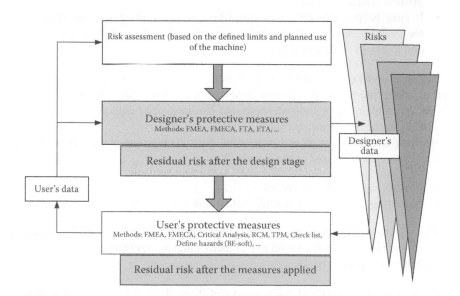

FIGURE 4.12 Procedures and methods of risk reduction.

4.3.1 SELECTION OF PROCEDURES AND METHODS FOR RISK ASSESSMENT

The theory and application of systematic risk assessment is a part of effective risk control processes. Methods for risk assessment or estimation allow the quantification of the risks and, thus, provide top managers with relevant information for risk reduction within the Man–Machine–Environment system. The following theses form the basis for the risk assessment methods selection (see also Chapter 1):

- In order to achieve a state of safety, the mere execution of the measures according to the regulations and standards will not suffice; it is also necessary to assess risk beyond the legal requirements.
- There is no zero risk, as there is no absolute safety.
- Safety is accepting a certain level of risk.
- The boundary for risk acceptability is not fixed; it varies according to the level of technical and cultural advances and possibilities in the field of science and technology, bearing in mind the fact that there is always space for improvement.
- Even risk assessment and the introduction of appropriate measures will never be a guarantee for non-occurrence of an injury, failure, or other undesirable events and, therefore, preventive measures must include the preparation for handling an accident.
- Staff members, users, and other personnel involved must be notified of residual risks.
- Risk assessment methods define the procedures that provide, on one hand, for complex and systematic assessment of whatever may cause harm, and on the other hand, for the focus on the most serious problems, which are the greatest sources of risks.
- It must be possible for the people who can cause the risk to manage it; for example, designers in their design, manufacturers in their products, employers in work they delegate.

Absolute understanding of these principles and their application in practice, as well as mastering the respective risk assessment methods, faces in practical life a variety of problems, especially in small and medium-sized enterprises at which no specialists in the field are employed.

Nowadays, there exist several procedures for risk assessment execution. In general, the procedures consist of these main steps:

- Identifying danger and hazard
- Assessing the degree of hazard, that is, risk assessment

The selection of a particular procedure for identifying dangers and hazards and for assessing risks depends on a variety of factors that are represented by the answers to the following questions:

- What is the goal of risk assessment (fulfilling the legal requirements and/ or risk reduction)?

- What is the time span within which first results are expected to be available?
- Is risk assessment a part of the audit or is it an individual procedure?
- Who and by what means will the risk assessment be executed?

The analyses used within the process of risk assessment are defined based upon:

- The amount of information to be collected and analyzed
- Time consumption

The following morphological analyses are used in the process of identifying dangers and hazards:

- Check-up sheets
- Analyses by means of format sheets
- Analysis by means of morphological fields
- Analyses by means of catalogues
- Equipment inspection in compliance with, for example, STN EN ISO 12100-1, 2, or STN EN ISO 14121-1

Risk is defined as a combination (function) of the probability of negative event occurrence and the consequence of a potential injury, health damage, or harm (R = risk, P = probability, C = consequence).

$$R = P \times C$$

Risk may be defined by

$$\text{a linear function:} \quad R = P \cdot C \tag{4.2}$$

or more precisely by

$$\text{a Cartesian product:} \quad R = P \times C \tag{4.3}$$

or by

$$\text{a nonlinear function:} \quad R = f(P, C) \tag{4.4}$$

During practical applications of risk assessment, it is at times advisable to use more parametric methods in a form of, for example, the extended definition of risk:

$$R = P \times C \times Ex \times O \tag{4.5}$$

where Ex stands for the duration of time during which the hazard affects the object in question, or exposition; and O stands for a possibility to use protective measures.

Risk is, however, influenced by far more factors than those included in the extended definition of risk. These factors are divided as follows:

- Measurable factors such as exposition time (Ex), speed of events occurrence (Sa), the number of people in danger (Np), the value of losses (Vl), and system parameters of weight (W), speed and acceleration (Sm), and height (H)
- Immeasurable factors such as possible danger identifiability (Ih), event occurrence observability (Im), operator's qualification (Q), human factor (Hf), environment influence (Ei), maintenance and inspections quality (Mc), complexity of system (Cs), and accident measures (Am)

The respective factors are to be considered in the process of determining the probability and consequence parameters, which makes their value more precise. The factors are represented by the functions:

$$P = f(P, Ex, Sa, W, Sm, Ih, Im, Q, Hf \ldots) \tag{4.6}$$

$$C = f(C, Np, Vl, W, H, Am, Im \ldots) \tag{4.7}$$

In addition, risk may be expressed as a function of several parameters that allow for the inclusion of the specific conditions of the Man–Machine–Environment system in which risk is understood as:

$$R = f(P, C, Ex, Sa, W, Sm, Mc, Ei, Ih, Np, Vl, Q, Hf \ldots) \tag{4.8}$$

In the process of risk assessment, the risk is estimated followed by its division into respective groups based on the risk intensity.

It is possible to effectively determine risk wherever there is a possibility to unequivocally and quantitatively assess the risk parameters, that is, the probability and consequence of a negative event; this includes technologies that archive detailed data on individual injuries, accidents, faults, and assess risks (hazards) that can be quantified by explicit values, for example, noise, vibration, dustiness, chemical substances content, and so on. In case of technical risks, it is possible on some occasions to define consequences by means of financial units, although this procedure is applicable only in case of incidents covered by insurance.

4.3.2 Risk Estimation Methods

The methods to be used in the process of risk assessment or estimation include:

- Risk matrix (currently most frequently used in the field of occupational health and safety)
- Risk graph
- Numeral point estimation of risks
- Quantified risk estimation
- Combined methods (mixed methods)

There are other methods to be found in special books that deal with the subject.

TABLE 4.1

Risk Matrix of the 6 × 4 Type

Occurrence Probability	Defined Frequency (per annum)	Consequence Severity			
		Catastrophic	Major	Great	Minor
Frequent	> 1	V	V	V	S
Probable	$1-10^{-1}$	V	V	S	N
Random	$10^{-1}-10^{-2}$	V	V	S	N
Low	$10^{-2}-10^{-4}$	V	S	S	N
Improbable	$10^{-4}-10^{-6}$	V	S	N	Z
Almost impossible	$< 10^{-6}$	S	S	Z	Z

Note: V = significantly high risk, S = medium risk, N = low risk, Z = negligible risk.

4.3.2.1 Risk Matrix

This tool for risk assessment offers a wide scope of applications thanks to its simplicity. It is based on the estimation of probability and consequence of an identified hazard. The main aim of the tool is to provide for the risk extent estimation, or provide information necessary for the risk assessment (see Table 4.1).

The drawbacks of this method include the fact that the parameters of the probability and consequences of a negative event are quantified by 'blunt' nonnumeric values, that is, values defined verbally. It is up to an assessor to select criteria for dividing risk into individual groups, which renders comparability of results defined by different assessors difficult and eventually biased.

4.3.2.2 Risk Graph

A risk graph represents a graphical output for risk estimation. It is based on a so-called decision tree in which every node represents a certain quantity, or risk parameter (e.g. consequence, probability of occurrence, the frequency of negative effect exposure, etc.), and the graph direction stands for a degree of severity (importance) of the given parameter.

Risk graphs are simple and graphically describe individual risk parameters, which contributes to the decision making on the means for risk reduction to a desirable level (see Figure 4.13).

With more than two branches, that is, if the given parameter consists of three options (e.g. probability of undesirable event occurrence P is defined by values $P1$, $P2$, and $P3$), the graph becomes more complicated and difficult to read. That is why, under these circumstances, a combination of a risk graph and a risk matrix is utilized (combined method). This procedure takes into account more possible parameters of hazard effects; however, the division into individual groups depends on the subjective evaluation by company's risk assessment specialist.

4.3.2.3 Numeral Point Estimation of Risks

This method is frequently used as a qualitative method for the initial selection of hazardous (dangerous) machines and machinery systems. The method makes use of

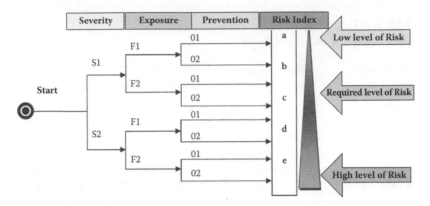

S   Injury/harm severity

S1   Minor (minor injury, reversible), e.g. scratch, cut bruise

S2   Minor injury (normally irreversible, including death), fracture, limb detachment, or
        crushing …

F   Frequency and/or duration of hazard effect (Exposure E)

F1   Two times or less per shift (random) or max. 15-minute exposure (short exposure)

F2   More than two times per shift or longer than 15-minute exposure

0   Possibility to prevent or reduce harm

01   Possible under certain conditions (e.g. the speed of the components is lower than
        0.25 m/s, staff members wear personal protective equipment …)

02   Impossible

FIGURE 4.13   Risk graph showing a desirable level of risk.

a point, or weight, expression for defining the importance or severity of risk parameters; it is a qualitative expression of 'risk levels' of the observed system or activity.
The final result is a combination of the final numerical values of the risk parameters,
for example, by summation or product.

    Description of the method's individual steps could be as follows:

- Probability parameter P of accident occurrence could be evaluated by points:
  - Very probable event occurrence $P \geq 100$ (almost certain)
  - Probable $99 \geq P \geq 70$ (may occur)
  - Improbable $69 \geq P \geq 30$ (low probability of event occurrence)
  - Very improbable $29 \geq P \geq 0$ (probability is close to zero)
- Consequence parameter C could be categorized as follows:
  - Catastrophic $C \geq 100$
  - Critical $99 \geq C \geq 90$
  - Medium $89 \geq C \geq 30$
  - Negligible $29 \geq C \geq 0$

    The final relation for risk estimation is defined as a sum of the probability and
consequence parameters as follows:

TABLE 4.2

Example of the Point Estimation and Assessment of Risk

Danger	Hazard	P	C	R/Degree	Risk Reduction	P	C	R/Degree
Moveable parts (punch, work-piece)	Contact with a moveable part	80	95	175/high	Protective cover	30	95	125/medium

$$R = S + C \qquad\qquad (4.9)$$

where R stands for the point estimation of system risk.

For an example of a dangerous activity assessment by means of the point risk estimation, see Table 4.2.

4.3.2.4 Assessment of Risk as a Bi-Parametric Quantity (e.g. MIL STD 882C)

Risk assessment by means of immeasurable quantification is used if there is no clear interface between the individual elements of the Man–Machine–Environment system, and in the process of the assessment it is difficult to precisely determine the values of probability (frequency) and consequence of a negative event. This includes MIL STD 882C elaborated by the US Department of Defense and is quite frequently utilized mainly by small and medium-sized enterprises.

The quantitative risk assessment is carried out on the basis of a matrix in which the relation between probability, consequence, and risk is clearly defined (see Table 4.3).

The above-mentioned point-based method defines four levels of risk within the range of 1 to 20 points. Determining the risk levels allows for the introduction of particular measures to reduce risk.

Table 4.4 includes an example of a particular risk assessment for a set of mechanical hazards. It is obvious that in the case of winding, the risk is unacceptable. The employer is supposed to take measures to make the risk levels fall back to the

TABLE 4.3

Risk Matrix

	Risk Matrix			
Probability/Consequence	I Catastrophic	II Critical	III Marginal	IV Negligible
A. Frequent	1	3	7	13
B. Probable	2	5	9	16
C. Occasional	4	6	11	18
D. Rare	8	10	14	19
E. Improbable	12	15	17	20

TABLE 4.4

Risk Level

Points	Risk Level
1–5	Unacceptable
6–9	Undesirable
10–17	Acceptable with inspections
18–20	Acceptable without inspections

acceptable risk group; it is also expected that inspections be carried out at regular intervals (or as a part of internal audits).

The point-based method is frequently used in practice to assess risks of individual types of machines and machinery. Versatility of the method is limited by the fact that the division of the risks into probability and consequence groups is based on a subjective assessment, which thus calls for a high-quality assessor.

4.4 LIFTING MACHINES AND RISK ANALYSIS

A crane is a lifting machine used to move a load vertically and horizontally within a designated area. The character of lifting machine operations represents a source of potential danger of a serious nature and, therefore, this kind of machinery is included in the group of restricted technical equipment.

On the basis of the Failure Mode and Effects Analysis (FMEA) method, it is possible to analyze hazards and, consequently, define risks that stem from a malfunction of the given equipment.

For example, the lifting mechanism of a travelling crane (crab; see Figure 4.14) performs the following working movements:

- Lifting the load with the suspension device
- Moving the load in 3-D space
- Handling the load on the suspension device (e.g. grab, ladle tipping)

The main parts of the lifting mechanism include:

- Electromotor
- Rope-winding drum or chain wheel
- Shafts and bearings of drums, pulleys, and wheels
- Rope and chain transmission pulleys
- Shafts, bearings, and other fittings of transmission pulleys
- Suspension components (ropes and chains)
- Geared transmission uncovered or located in a box
- Couplings
- Brakes
- Mechanism to control the limit switch, a pulley block with a hook

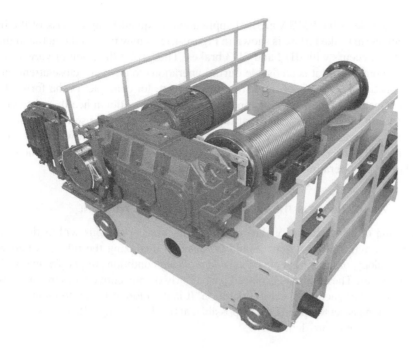

FIGURE 4.14 Lifting mechanism of a travelling crane—crab.

In order to identify dangers, analyses are done for the following:

- Brake electromotor
- Coupling between the electromotor shaft and the gearbox
- Gearbox
- Coupling at the gearbox output
- Rope drum, speed limiter, load limiter
- Steel rope
- Binding device

Based on the detailed analyses, including the statistics on failures and accidents during the operation of the lifting machine, the following types of hazards have been identified:

- Brake failure—renders the brake unusable (e.g. pressure spring cracked)
- Brake lining excessively worn out
- Failure of the electromotor rotor bearings
- Mechanical failure of the functional connecting parts
- Gear damage
- Failure of the load limiter
- Wire breakage
- Knots, loops, deformations, crimped rope, basket deformation of a rope
- Inappropriate and damaged suspension device

An example of the FMEA method application for quantifying the risk of the lifting mechanism brake failure is shown in Figure 4.15; it may be stated that the highest level of risk occurs with d112 and d113 brakes. Thus, from the point of view of this travelling crane user, it is advisable during operations to pay very close attention to these functional elements of the lifting mechanism, for example, in the form of an effective method of maintenance control. As far as occupational health and safety is concerned, it may be concluded that the highest risk will be caused by the brakes' failure. It is therefore necessary to ensure that no person is present within the crane working area during load lifting. This requirement is normally included in the safety regulations for operating this kind of lifting machine.

4.4.1 Application for the Workplace of a Lifting Machine

Evaluation of a human factor during lifting machine operations within the Man–Machine–Environment system should be done by analyzing the relations between the technology, organization of works, and working conditions of people involved in the operations. The emphasis is placed particularly on preventive measures to reduce risks before the machine is set to operation. It is therefore efficient to identify risks before a crane is set to operation—if possible, at the design stage. Regarding this, the following priorities are defined:

- Elimination or reduction of dangers by means of a safe crane design
- Using protective measures in case the dangers were not eliminated by the safe design of the crane
- Notifying a user of residual hazards that may occur during crane operation

The outcomes of the hazards analyses and risk assessment are the basis for a company's management in a decision-making process of selecting effective methods and procedures to reduce risk. Both lifting machine manufacturers and users may contribute to risk reduction (see Table 4.5).

The government control authorities issue various brochures and manuals for users of different machinery. Within the European market, designers are given access to a number of manuals that include all the regulations and standards. The manuals are updated annually to include the latest enactments, for example, the manual by the International Social Security Association (ISSA) Section for Machine and Machinery Safety, Mannheim/Germany. In Table 4.6 and Table 4.7 are some of the mechanical and thermal hazards during operation of abridge crane as an example.

4.4.2 Risk Assessment at the Workplace Using the Complex Method

The workplace risk assessment method was developed by ISSA's Section for Machine and Machinery Safety. When applying this method, it is necessary to take into consideration the elements of the analyzed system that can be ignored and the elements to which close attention must be paid. A system, within this method, is defined as a set of elements that contribute to a certain activity. Common systems in which human risks exist include a human factor (abilities) that functions in a certain

Organization									Processed on:			XI.00	
Equipment: Traveling crane 250t Segment: **Lifting mechanism/crab 250t**			Identification of Dangers, Hazards, and Risks						Checked on: Page: Author: TU SjF, Košice				
									SU				
Code	Danger	Hazard (Form of Failure)	Consequences	Reasons (for Failure)	Checks	VZ	VY	OD	MR/P	Q	E	S	Notes
d111	Disc brake BKD 630—disc	Loosened	Repair	Wrong positioning of engine and gearbox	1 × per week	3	3	5	45	N	N	N	
d112	Disc brake BKD 630—disc	Holes in disc	Repair	Wear	1 × per week	6	5	5	150	N	N	N	
d113	Disc brake BKD 630—coupling	Holes	Repair	Wrong positioning and screws loosened	1 × per week	10	4	5	200	N	N	A	Maintenance—limited space
d114	Disc brake BKD 630—pins and castings	Noise	Repair	Wear	1 × per week	6	3	5	90	N	N	N	

FIGURE 4.15 FMEA method for the lifting mechanism brake.

TABLE 4.5

Example of Risk Assessment

Hazards Group	Type of Hazard	Danger	Related Regulations for the Main Type of Danger	Probability	Consequence	Risk
Mechanical	Cutting	Knife	xxxxxxxxxx	B	IV	16
Mechanical	Winding up	Winch	xxxxxxxxxx	A	III	7

TABLE 4.6

Selected Responsibilities of Manufacturers and Users of Machines and Machinery Systems

Manufacturer

Safe design

Own/internal/safety

Technical protective measures

User's manual with information on residual hazards or dangers

User

Education, trainings

Personal protective equipment

Organizational measures

Residual risk—acceptable risk

Detailed and constant updating

working process and uses special tools and aids. The main principle of this method is based on appropriate assignment of points to the individual system elements and on defining acceptable risk. The most problematic area in the process of risk assessment is the evaluation of the human factor. In order to reduce a subjective factor, it is advisable for the evaluation to be carried out by the same person or team of experts.

Points are assigned to a particular risk that exists in a working process and is a function of the individual system element parameters. The points assigned provide the final risk evaluation.

TABLE 4.7

Selected Hazard Types

Division plant:	Steel mill		
Equipment:	Cabin-operated traveling crane		Sheet No.:
Associated system:	Convertor		Date:
Profession/ Position:	CRANE OPERATOR	Compiled by:	SjF TU KOŠICE

Hazards	Dangers	Risk Description
Mechanical	• Crushing, smashing	• Moving on and around the crane during and after operations—handrails, safety switches • Climbing the crane—ladder protections, blocking device • Load oscillation during crane's starting and braking, diagonal pull • Staying in the load-handling area—training and practice • Handling the suspended load
	• Cutting, shearing	• Steel rope inspection and maintenance • Sharp edges
	• Winding	• Electro motors, winding mechanism—protective shields, warning signs, drums, pulley block
Thermal	• Burnt by flame	• Fire during operation
	• Burnt by radiation	• During fire or in case of liquid metal splashing
	• Burns caused by splashing	• Liquid metal splashing

4.4.2.1 Assessment of Risk Caused by Equipment

Suggested Evaluation

1. Determination of potential harm

Dangerous injuries with minor consequences
(sprains, contusion, minor cuts)   1

Dangerous injuries with major consequences
(fractures, deep cuts)

Dangerous injuries with permanent consequences
(death)   10 $S =$ ☐

2. Danger of exposure (frequency and duration)

Temporary minor exposure (automatic machines   1
working well, rare interventions …)

Exposure frequently reappearing (hands affected in
the working cycle)

Frequent or constant exposure (manual activity, e.g.   2
tool replacement) $Ex =$ ☐

Suggested Evaluation

3. Probability of dangerous situation occurrence (connected to the equipment factor)

Low (lack of dangerous elements, reliable, practical
and safe protective equipment, safe switch-off)　　　　0.5

Medium (complete protective equipment, in good
conditions yet impractical, so many activities done
with no protective equipment)

High (missing or insufficient protective equipment,　　　　1.5
possible dangerous impacts on a running machine)　　　　　　　$Wa = $ ☐

4. Skoda prevention and reduction possibilities

Major (warning the staff prevents harm)　　　　0.5

Minor (danger effect mechanism is sudden and
unexpected)　　　　　　　　　　　　　　　　　$Ve = $ ☐
　　　　　　　　　　　　　　　　1

Final evaluation of the equipment factor: $M = $ ☐

$$M = S \times Ex \times Wa \times Ve \tag{4.10}$$

The following steps represent the working process risk assessment:

- Evaluation of the item total risk
- Evaluation of the environment effect
- Evaluation of a person's ability to handle risk
- Final risk calculation
- Comparison of the calculated risk and the acceptable level of risk
- Introducing the measures

4.4.2.2　Evaluation of the Environment Effect

Suggested Evaluation　　　**Final Result**

1. Workplace and impact zones layout

At a single level　　　　0.5

At several fixed levels

Using accessories and aids (ladder, foot step, ...)

Clear and spacious walkways

Narrow and insufficient walkways　　　　1
　　　　　　　　　　　　　　　　　$Ua = $ ☐

2. Working environment

Insufficient illumination　　　　0.3

Nondisturbing noise, pleasant environment (acoustic
signals well absorbed, cabin air conditioned)

Disturbing noise (acoustic signals absorbed
insufficiently)

Unpleasant environment (high temperature, dust,
humidity, draft)

Disturbing onerous environment · 0.6

$Ub =$ ☐

3. Other obstacles

Ergonomically suitable layout of controls, screens, · · · · · · · · 0.2
displays, data, and material supply

Unsuitable layout of controls, screens, displays, data
and material, supply

Minor physical workload (lifting and carrying loads)

Major physical workload (lifting and carrying loads) · · · · · · · 0.4

$Uc =$ ☐

Final evaluation of the equipment factor: U = ☐

$$U = Ua + Ub + Uc \tag{4.11}$$

Individual points assigned in the tables above are merely suggestions; an assessor is free to create a more detailed scale of points' assignment. The boundary point values must, however, be left untouched as per the table.

Provided that $R_A = 15$ is the agreed value of acceptable risk for the analyzed type of profession for a traveling crane, it may be assumed that the crane operator's profession is not one that requires an immediate introduction of intensive measures for risk reduction.

4.4.2.3 Person's Ability to Handle Risk

Suggested Evaluation   Final Result

1. Person's qualifications

Specially trained and educated person with · · · · · · · · · · · · · · · · · 10
experience

Specially trained, educated, or experienced person

Person is not specially trained, educated, or · · · · · · · · · · · · · · · · 0
experienced

$Q =$ ☐

2. Physical and mental factors

Person is in good physical condition for the · · · · · · · · · · · · · · · · · 3
assigned work

Person is not in sufficient physical condition for the · · · · · · · · 0
assigned work

$\phi =$

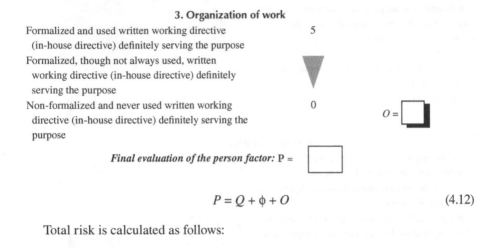

3. Organization of work

Formalized and used written working directive (in-house directive) definitely serving the purpose	5
Formalized, though not always used, written working directive (in-house directive) definitely serving the purpose	
Non-formalized and never used written working directive (in-house directive) definitely serving the purpose	0

$O = \Box$

Final evaluation of the person factor: P = $\Box$

$$P = Q + \phi + O \qquad (4.12)$$

Total risk is calculated as follows:

$$R = M \cdot U - P \cdot (M/30) \qquad (4.13)$$

Figure 4.16 shows an example of risk assessment by means of a multiparametric method (ten identifiers) for the lifting machine operational staff—a travelling crane operator.

The working conditions on a travelling crane are specific with respect to performing risk analyses for all the jobs involved in the crane's working process. Considering the nature of the crane's activities, risk analyses are necessary for a crane operator's profession and, moreover, for a slinger, and eventually for a spotter (taking into account difficult conditions for working with metallurgy technologies).

Although some countries' legislation does not recognize the term *acceptable risk* (which is seen as an imperfection), on many occasions the term *residual hazard* is used, supplemented by directives that instruct on measures for the elimination and reduction of the hazard. In practical life, it is possible to eliminate the hazard (risk) only in a limited number of specific cases (e.g. cranes excluded from the technological cycle, load oscillation eliminated by fixed suspension). From a practical point of view, it is useful to apply *reduction* or *minimization* of hazards (risk) and the final value of the reduction or minimization must be defined by 'acceptable' hazard (risk). The level of the risk is determined by the equipment (travelling crane) user based on the levels of risk he wishes to achieve.

Risks have been assessed for all the jobs involved in execution of active working operations of a travelling crane within the metallurgical production technology (steel mill). The technological cycle assessed includes handling operations by means of a steel-works crane controlled by crane operators, located in a crane cabin, and by crane spotters operating within the handling area close to the load, mostly ladles filled with liquid iron. The outcomes of the assessment are shown in Table 4.8.

It is obvious from Table 4.8 that the spotter's job is significantly more dangerous than the crane operator's position of working in a cabin. The risk level of the charning crane operator's job is higher than that of a casting crane operator because of the hazard of liquid iron splashing when filling the convertor. The outcomes gained from

Date:		Plant:		
System:	Steel-works crane	Assessed by:		
Profession:	Crane operator	Name:		
Risk Parameters		Label	Final Result	Value Interval
Effect of Equipment				
Stating probable harms	Major injury, in a critical case followed by death.	S	7	1–10
Duration/exposure/hazard	Handling the liquids is interrupted, not continuous.	Ex	1.1	1–2
Probability of hazard occurrence	Based on the experience the probability is not very high.	Wa	0.7	0.5–1.5
Disrupting the failure mechanism	After the occurrence of an undesirable event it is not possible to disrupt it.	Ve	1	0.5–1
$M = S.\ Ex.\ Wa.\ Ve =$			5.39	0.25–30
Effect of Environment				
Workplace layout	Since it is a conventional cabin, the requirements are met only partially.	Ua	0.7	0.5–1
Workplace environment	The cabin is not air conditioned. Significant effects of high temperature, dust, and exhalation.	Ub	0.5	0.3–0.6
Other obstacles	Cabin oscillations (vibrations), noise	Uc	0.3	0.2–0.4
$U = Ua + Ub + Uc =$			1.5	1–2
Effect of Operators				
Operators' qualifications	Operator's qualifications as required by directives; insufficient experience	Q	8	0–10
Psychological factors	Operator's job is stressful; handling dangerous substances	φ	2	0–3
Organization of works	Not all of the operator's manuals and instructions are available.	O	3	0–5
$P = Q + \phi + O =$			13	0–18
Risk Value				
$R = M.\ U - P.(M/30) =$			5.749	0–60
Risk assessment:	**Acceptable risk:**		5.749 <15	

Note: When estimating the risk parameters, possible inappropriate working conditions are taken into account, as defined by a team of experts following their workplace observations.

FIGURE 4.16 Specific example of assessing a crane operator's profession.

the evaluation of injuries at the analyzed operations support this assumption. Using a 'humanized' cabin, it is possible to alter the Ub (0,5 to 0,3) and Uc (0,3 to 0,2) parameters, and reduce the total risk provided the given plant's management considers the level of operator's risk, as stated in Figure 4.16, as unacceptable. The reduction of the spotter's risk in the given working conditions is plausible by means of organizational measures, that is, reducing the exposure time Ex and decreasing the probability of dangerous injury occurrence Wa, Ua, Ub by modifying the place from which the activities are performed as well as the type of data transfer to the crane operator.

TABLE 4.8

Risk Comparison at Travelling Crane Workplace

Profession	S	Ex	Wa	Ve	M	Ua	Ub	Uc	U	Q	Fi	O	P	Risk
Liquid steel spotter	10	1.5	1.2	1	18	0.8	0.6	0.3	1.7	8	1.5	3	13	23.1
Charning crane operator	10	1.1	0.7	1	7.7	0.7	0.5	0.3	1.5	8	2	3	13	8.213
Casting crane operator 240t	7	1.1	0.7	1	5.39	0.7	0.5	0.3	1.5	8	2	3	13	5.749
Casting crane operator 220t	7	1.1	0.7	1	5.39	0.7	0.5	0.3	1.5	8	2	3	13	5.749

4.5   PARTIAL CONCLUSION

Risk assessment methods play a crucial role in risk management. It is important to utilize unified terminology in debates and discussions so that all parties involved have a unified notion of the individual stages of the causal relation of a negative event occurrence. For the risk reduction strategy, it is crucial to make use of effective methods for interrupting the relationships because management's decision making on this process is plausible solely on the condition that relevant data on the risk levels of an observed machine or machinery system are available. Methods for technical equipment risk assessment can be used for machines of various types of complexity and potential harm. The selection of the procedures and methods for risk assessment depend on the goals to be achieved. Each procedure may lead to effective results if they are executed by a team of experts consisting of specialists in the field of occupational health and safety, experts from the analyzed plants, and staff representatives. The selection algorithm, that is, the qualitative estimation, within which the quantitative methods for more accurate risk estimation and assessment are introduced in case of more serious risks, has proven successful on many occasions.

It has to be taken into consideration, however, that outcomes of the analysis must create conditions for interrupting the casual relationship of negative event occurrence, that is, for elimination or, more likely, reduction of risk. When applying risk analysis in the machine design stage, the analysis outcome must enable a designer to implement the measures for risk reduction at the project stage and, as such, to interrupt the causal relationship in its early stages. If this is not plausible or economical, residual risks are to be included in the technical terms and conditions or in the manual, and the designer will propose measures to be applied in the process of setting a machine to operation.

It is then the task of the employer to apply these measures, including the systematic inspection of the staff members applying the measures in their everyday working activities.

BIBLIOGRAPHY

Karwowski, W., and Marras, W.S. 'Risk assessment and safety management in industry', in *The Occupational Ergonomics Handbook*, Boca Raton, FL: CRC Press LLC, 1998, pp. 1917–1948, ISBN 0-8493-2641-9.
Karwowski, W., and Salvendy, G. *Advances in Human Factors, Ergonomics, and Safety in Manufacturing and Service Industries,* Boca Raton, FL: CRC Press, 2011, p. 826.
Pačaiová, H., Sinay, J., and Glatz, J, *Bezpečnosť a riziká technických systémov,* edited by SjF TUKE Košice, Vienala Košice 2009, ISBN 978-80-553-0180-8.
Sinay, J. 'Konkrétny príklad posúdenia rizika pri prevádzke zdvíhacieho stroja', Conference: Lifting Machines in Theory and Practice, VUT Brno, May 1999, pp. 60–66, ISBN 80-214-1329-8.
Sinay, J. 'Rizika pri prevádzke zdvíhacích strojov', (New Findings on Lifting Machines: Risks), ČSMM-L-OSZZ-Prague, June 9, 2010, p. 17. Published in *Journal of Česká společnost pro manipulaci s materiálem*, Prague, CR, December 2010.

Sinay, J. 'Einige Überlegungen zur Risikoanalyse während des Kranbetriebes', in *Der Kran und sein Umfeld in Industrie und Logistik*, 19 Internationale Kranfachtagung Magdeburg: 31 März 2011, Magdeburg. Magdeburg, ILM, 2011, pp. 119–125, ISBN 13:978-3-930385-74-4.

Sinay, J., and Badida, M. 'Quatifizierung der Risiken beim Kranbetrieb', *F+H Fördern und Heben* 49/Nr. 4, Mainz, GER, 1999, pp. 273–276, ISSN 0341-2636.

Sinay, J., Badida, M., and Oravec, M. 'Anwendung der Mehrparameter-Methode zur Risiko-Beurteilung im Rahmen des Risikomanagements', *Technische Uberwachung-TU Nr*, VDI-Verlag, Düsseldorf, April 1999, pp. 51–54, ISSN 1434-9728.

Sinay, J., et al. *Riziká technických zariadení: Manažérstvo rizika*, OTA Košice, 1997; also as a CD, ISBN 80-967783-0-7.

Sinay, J., Kotianová, Z., and Pačaiová, H. 'Posudzovanie rizík technických zariadení: Postupy a metódy', in *Occupational Health and Safety*, 2009, Ostrava: VŠB-TU, 2009, pp. 263–270, ISBN 9788024820101.

Sinay, J., and Laboš, J. 'Manažment rizika počas technického života produktu: Potreba alebo samozrejmosť', *Bezpečná práca*, January 2003, pp. 15–17, ISSN 0322-6347.

Sinay, J., and Majer, I. 'Human factor as a significant aspect in risk prevention', 2nd International Conference of Applied Human Factors and Ergonomics, 12th International Conference on Human Aspects of Advanced Manufacturing – HAAMAHA, July 14–17, 2008, Las Vegas, Nevada, USA, Session 20.

Sinay, J., and Nagyova, A. 'Causal relation of negative event occurrence: Injury and/or failure', in *Advances Factors, Ergonomics, and Safety in Manufacturing and Service Industries*, AHFE Conference 2010, Boca Raton, FL: CRC Press, 2010, pp. 818–827, ISBN 978-1-4398-3499-2.

Sinay, J., and Oravec, M. 'Viacparametrické metódy klasifikácie rizika', Conference: Topical Issues of Work Safety, 11th International Conference, VVUBP Bratislava, 1998, pp. 64–72.

Sinay, J., Oravec, M., Majer, I., and Sloboda, A. 'Methods of risk evaluation', 3rd International Conference Globalna varnost, Bled, Slovenia, June 1998, pp. 17–21.

Sinay, J., Oravec, M., Pačaiová, H., and Tomková, M. 'Application of technical risk theory for evaluation of gearboxes damaging processes', *International Symposium, From Experience to Innovation–IEA 97*, Tampere, 1997, pp. 560–562, ISBN 951-802-197-X.

Sinay, J., and Pačaiová, H. 'Analyse und Bestimmung der Risiken im Hubwerk eines Hüttenkranes', in Kranautomatisierung Komponenten Sicherheit im Einsatz, Magdeburg: LOGiSCH, 2002, pp. 31–42, ISBN 3930385376.

Sinay, J., and Pačaiová, H. 'Integrierte Einstellung zur Beurteilung des Risikos von Hebezeugen', in 13 Internationale Fachtagung 2005, *Von der Automatisierung bis zur Zertifizierung*, Magdeburg, IFSL Otto-von Guericke Universität Magdeburg, Reihe III: Tagungsberichte Nr. 202, June 2005, pp. 147–162, ISBN 3-9303385-53-8.

Sinay, J., and Pačaiová, H. 'Integrierte Verfahren zur Beurteilung der Risiken bei Hebemaschinen', in *Von der Automatisierung bis zur Zertifizierung*, Magdeburg, IFSL, 2005, pp. 149–159, ISBN 3930385538.

Sinay, J., Pačaiová, H., and Oravec, M. 'Application of risk theory in Man–Machine–Environment systems', in *Fundamentals and Assessment Tools for Occupational Ergonomics*, Boca Raton, FL: CRC Press, Taylor & Francis, 2006, pp. 8-1–8-11, ISBN 0849319374.

5 Certain Risks and Principles of Their Management

5.1 NEW AND EMERGING RISKS

New and emerging risks represent a topical issue in the area of occupational health and safety. New and emerging risks appear in locations in which new machines and machinery are applied within new technologies, materials, and processes, or under the influence of various factors stemming from corporate social changes, that is, where new forms of organization of work are applied, for instance:

- Nanotechnologies, nanoparticles, and very discrete particles
- Mechatronic systems (e.g. modern automobiles, robots, automated cranes)
- Biotechnologies
- Renewable energy sources
- Linear production processes
- Time-limited work agreements
- Aging employees
- Growing work intensity
- Using Information and Communication Technologies (ICT) techniques (hardware + software) and technologies
- Combined load on muscle-skeleton body system and psycho-social risk factors (e.g. uncertain future of people and families), general uncertainties

A new risk is a risk that:

- has not yet existed and is caused by new processes, technologies, workplaces, and organizational or social changes, or
- is a long-term problem reconsidered to be a risk as a result of changes in the social or public perception, or
- a long-term unsolved problem that is re-evaluated as a risk based on new scientific discoveries.

Risk occurs as a result of a variety of changes provided that:

- the number of dangers (hazards) contributing to the risk is growing, or
- the period of exposure to hazards contributing to the risk is getting longer, or

- the effect of the hazard on employee health is intensifying (seriousness of the effects and/or a number of people affected).

The identification and control of new and emerging risks as a part of risk management are subjects of a number of international research projects. Requirements that must be applied in the process of managing these risks determine both the content of educational systems (curricula) and the process of obtaining respective qualifications. Taking into account labour market globalization and the fact that safety goes beyond the borders of countries, the subjects of prevention and minimization of risks within the European Union (EU) have become a part of international research projects of the highest priority, such as iNTteg - Risk CP-IP213345-2, 'Early Recognition, Monitoring and Integrated Management of Emerging, New Technology Related Risks,' within the 7th Framework Programme European, which includes sixty-nine institutions of eighteen EU countries coming from different research fields: universities, research institutes, and public institutions dealing with the research and application of risk management methods, including the application of appropriate ergonomic solutions as a means of effective prevention.

Organizations must be able to identify new and emerging risks and to propose sufficient measures for their elimination or reduction to an acceptable level, respectively. An aging labour force and the changes it brings represent one of the new risk factors. Solutions in this area will cause changes in both the work environment and work life quality.

In order to achieve effective new and emerging risk control within the modern methods of occupational health and safety (OHS) management, including effective prevention, and within the integrated management systems (quality, safety, and environment), it is absolutely necessary to provide all the parties involved in the working process with new knowledge and skills from all relevant technological fields, such as ICT, natural sciences (chemistry, physics, biology, ergonomics), and humanities (sociology, psychology, political science), using new forms and methods of education.

It is obvious that an expert in the field of safety cannot replace an engineer, electrician, physicist, chemist, psychologist, ergonomist, doctor, sociologist, and other expert in a different field of expertise. The safety expert's role is, most of all, to apply his/her knowledge and skills in the process of analyzing and assessing risks as a part of teamwork. Current requirements expected from experts of various technical fields include the integration of new areas of science and research into the educational process, for example, knowledge of informatics, mechatronics, system technology, and environment technology.

Preventing new and emerging risks became one of the main goals included in the European Community Development is Horizon 2020 (2014–2020). The issue is dealt with by a number of projects and studies in the area of occupational health and safety.

The current strategy of the Community for 2007–2012 calls for the continuation of the given direction and for the improvement of procedures in risk management, including the risks related to new technologies, biological risks, complex risks, risks within the Man–Machine–Environment context, and demography-induced risks.

5.2 IMPACT OF INCREASING WORKING AGE ON OHS AS A NEW RISK

The field of occupational health and safety is an ever-developing area responding to new information, technological advances, or social situations. Regarding this development, new approaches to creating conditions for effective management of processes in occupational health protection and increasing occupational safety are required.

Nowadays, the fast-paced life style and never-ending advance of new technologies are commonplace, which is caused by the effort to produce faster, more efficiently, at higher quality, cheaper, and at low costs. It is undeniable that technological development influences various areas of society. Particular technological change, combined with the effects of other marginal factors, systematically generate changes in a variety of fields. For example, the development of safety technologies combined with other factors results in reductions in a number of serious injuries, fatal injuries, and in man's improved health conditions. This very fact, combined with the decrease in numbers of newborns, leads to an aging of population.

The aging of the active labour force could be included in the group of emerging risks. Unlike past years, when people would retire from active working life as early as in their fifties, today it is necessary for people to work until an older age in relation to the sustainable development of the world's economy. The changes in working abilities at an older age will, therefore, be even more obvious and new effective solutions to provide this group of employees with a healthy and safe working environment will have to be sought. This fact is further treated in the Slovak Republic by Act No. 124/2006's Article 5(2)(f), which defines an employer's obligation to 'adapt the work to worker's abilities and technological advances.' Paragraph g of the same Act defines a worker's duty to 'take into consideration people's abilities, potentials and features of the character when designing a workplace, selecting the working tools, and the working and production processes in order to eliminate or reduce the effects of the harmful factors of hard work and monotonous work on an employees' health conditions.'

The EU finds itself in a process of significant population aging. According to the latest Eurostat estimations, issued in 2008, in the EU by 2060 there will be only two people of a productive age (15–64 years old) for each person over 65 years of age, as compared to the current ratio of four people of a productive age for each person over 65 years of age.

The reason for the population's aging process is the increasing proportion of people 60 years of age in the overall population, with the portion exceeding twenty percent in the EU member states as well as in many other countries. The statistics show different pace and progress of aging in individual countries, however, the presumed change is significant and apparent in Figure 5.1. The *baby-boom* period, typical for its high number of newborns, is followed by the current demographic conditions defined as the *grandpa-crack*.

A significant opportunity to solve the issue of the demographical aging and maintain intergenerational solidarity is represented by efforts to ensure that the generation of the high-birth-rate period remains in the labour market and stays healthy, active, and independent as long as possible [the EU's material KOM(2010) 462 Final].

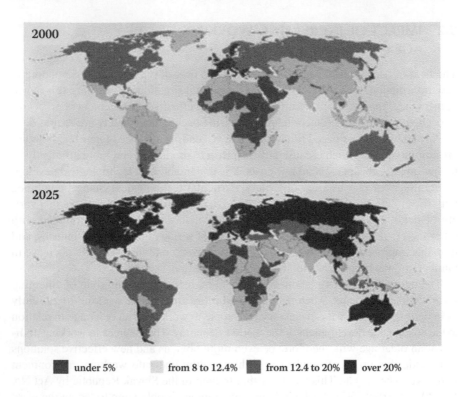

FIGURE 5.1   Worldwide aging levels.

Within the employment strategy, the member states implemented measures to stop the early retirement trend, which resulted in the EU-27's 55–64 age group unemployment rate rising from 36.9 percent in 2000 to 46 percent in 2009.

In light of the above-stated factors in relation to the labour market, it follows that older employees will constitute the fastest-growing labour force segment.

5.2.1   OLDER WORKER: DEFINITION

The United Nations Organization has suggested that 60 years of age be considered the boundary of old age. This suggestion, however, merely copies the general practice of taking retirement at this age rather than based on biological reasons. From the point of view of occupational medicine, symptoms of aging appear as early as age 45, with a decrease in the functional capacity of the body with regard to workload. Under these circumstances, people over 45 years of age could be considered aging employees.

The number of people who will fall into the category of older employees will be increasing in the years to come, as proven by the current demographic forecasts. Health conditions, changes in abilities and capabilities, and working capacity vary with growing age, but for the sake of a sustainable pension and the social welfare system, the retirement age in the EU has constantly been increasing.

According to the EU, those legislative measures that increase the working age will play a major role in this area and their existence is absolutely essential.

However, the increased retirement age, apart from positive effects, also has negative aspects, especially in the field of occupational health and safety.

It is both in employers' and the public interest to look for ways to create a safe working environment for this segment of employees and create conditions for so-called healthy aging in compliance with the World Health Organization's strategy.

5.2.2 AGING LABOUR FORCE AND CHANGES IN WORKING ABILITIES

Physical and mental abilities form a basis for man's working abilities. Studies dealing with working ability have proven that abilities decrease with increases in the age of workers. The reason is the loss of physical strength and health limitations, which result in a lower physical working capacity (see Figure 5.2).

Working ability is influenced mainly by the following working environment factors:

- Job description/physical workload, for example, static muscle load, lifting and moving loads, sudden load increase, repeated inappropriate and unacceptable moves
- Stressful and dangerous working environment, for example, noise, smoke, heat, and humidity
- Working environment layout, for example, conflicts, inappropriate planning and inspection, fear of failure and making mistakes at work, time pressure, lack of appreciation, and lack of freedom in decision making and personal development

The aging process is characterized by certain physiological and psychological changes that will be of significant importance in relation to employing the older generation.

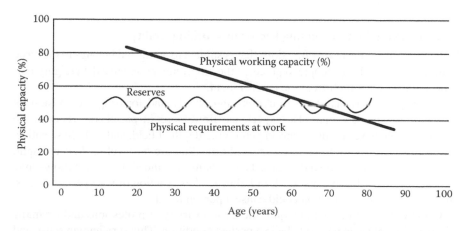

FIGURE 5.2 Relationship between work requirements and worker abilities.

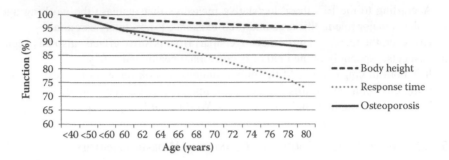

FIGURE 5.3 Function versus age.

As far as the physiological aspect is concerned, certain symptoms of aging can be observed as early as at the age of 45, but with major individual discrepancies. In general, the following changes could be included on the aging symptoms list:

• Changes in the functioning of the senses (e.g. sight, hearing)
• Changes in neuromusculoskeletal functions concerning movement (joint and bone functionality, muscle functionality, movement functions)
• Changes in sense functions
• Changes in mental functions
• Changes in cognitive functions

The natural aging process of a human being brings about some 'limitations' in body functions, that is, the same working conditions create greater load for the organism.

Progress and symptoms of aging are highly individual and depend on a variety of factors, such as sex, genetic predispositions, lifestyle, standard of living, eating habits, and their combinations. Generally speaking, these symptoms mostly occur between the ages of 40 and 50, or later, depending on the kind of change (see Figures 5.3 and 5.4).

5.2.2.1 Eyesight as a Limiting Factor in Working Ability

Eyesight is one of the bodily functions that is subject to aging and belongs to the basic human senses enabling people to perceive the environment, receive data (eighty percent of data) and stimuli from the environment.

Older employees may have trouble seeing clearly and focusing their vision on a distant object; they may also experience limitations in peripheral vision, have blurred vision, have complications in perceiving depth of field, and be light sensitive. These people are also prone to other conditions that are harmful to their eyesight, such as cataracts and retinal diseases. These changes in the ability to see things may increase the risk of accidents. Lack of balance, slow reaction times, eyesight issues, and insufficient concentration could cause a person to fall.

Good vision is an essential requirement for a variety of professions and for many of those, eyesight is expected to be in a perfect condition. This very human sense and its changes could influence the quality of both work and life. The complexity of the

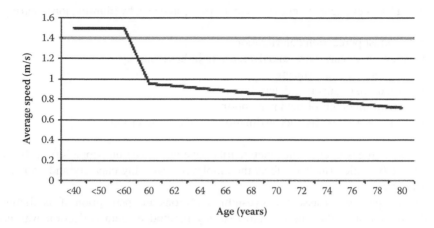

FIGURE 5.4   Average speed versus age.

visual apparatus (see Figure 5.5) is the reason why this organ is subject to various external factors and to internal changes of the organism.

There are a number of reasons why eyesight deteriorates:

- Change in eye's refined adaptability
- Lens hardening
- Atrophy of lens-thickness adapting muscles
- Changes in eye's light sensibility

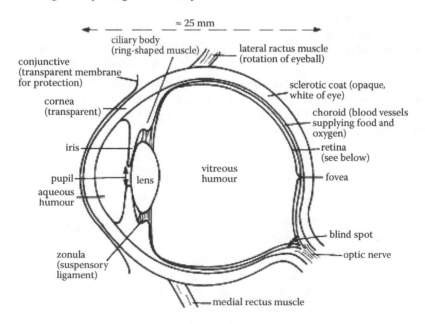

FIGURE 5.5   Visual apparatus.

This is a list of eyesight changes that can be prevented (e.g. by illumination resetting):

- Contrast perception deterioration
- View angle narrowing and decrease in light sensitivity
- Tendency to fixed staring
- Colour distinguishing deterioration
- Change in depth-of-field perception
- Deterioration in distance estimation and dynamic vision

These changes in eyesight functionality have an impact on people's overall sensor capacity, which further affects the employee's working capacity and ability to avoid dangers.

For the purpose of assessing eyesight conditions and perception of its changes with growing age, the questionnaire survey method of data collection was utilized within a group of respondents between 23 and 61 years of age working for a selected company.

The survey on the effects of the working environment on eyesight included responses from fifty-two respondents from the Košice region. The respondents anonymously answered questions about age, sex, profession, eyesight condition, and the effects of the working environment on their eyesight. Moreover, they stated their opinions on modifications of the working environment for employees over 50. The respondents are divided into the following categories:

20–30 years of age
30–40 years of age
40–50 years of age
50–60 years of age
60+ years of age

It follows from the survey responses that good eyesight conditions were reported by the respondents of the 20–30 and 30–40 years-of-age categories; the older age categories have admitted deterioration of their eyesight with a high percentage of prescription glasses used (Figure 5.6).

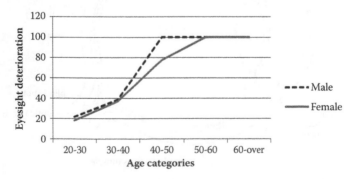

FIGURE 5.6   Eyesight deterioration (%) in relation to age.

Deterioration of eyesight has been stated by as much as seventy-nine percent of the 40–50 age group respondents. According to the survey, deterioration of eyesight occurs in respondents over 40 years of age. All the respondents from 50–60 and 60+ age categories have confirmed using prescription glasses. The findings may be biased by the number of respondents and their job description, since ninety-four percent stated that good eyesight conditions are essential for performing their work duties.

The questions used in the survey included the following:

1. Is eyesight and its good conditions essential for performing your work duties?
2. Have you noticed eyesight deterioration at your older age?
3. Do you think that the modification of your working environment (lightning, signalization, monitor contrast, altering eyesight-harming activities) would have a positive effect on your comfort?
4. Do you think that modification of the working environment for employees over 50 is necessary due to age-induced changes?
5. Do your eyes feel tired during and after your workday?

The individual responses are shown in Figures 5.7 through 5.10. The responses from the 60+ age group are not included due to a low number of respondents in this category. Figure 5.11 shows the responses from all categories.

The increasing number of older employees is obvious in all professions and in all positions. The trend of a growing number of older employees is expected to occur also in managerial positions. The knowledge, experience, and skills of older employees are the reason for their involvement in the active working life beyond retirement age. Furthermore, working on these positions causes an overload of the visual apparatus, since a majority of managerial positions, if not all, include working with computer display units, which cause major eye strain to eyesight.

Both legislators and employers should take into account the related issues and introduce measures to provide a safe and healthy working environment with respect to all employee groups. Currently, the major focus of the professionals in the field

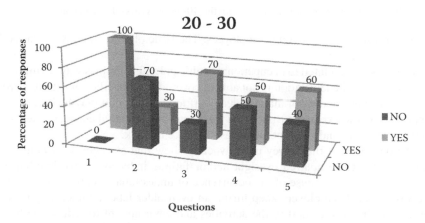

FIGURE 5.7   Percentage of responses from the 20- to 30-years-of-age category.

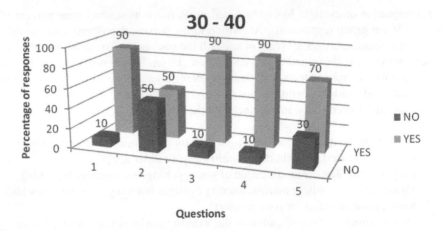

FIGURE 5.8   Percentage of responses from the 30- to 40-years-of-age category.

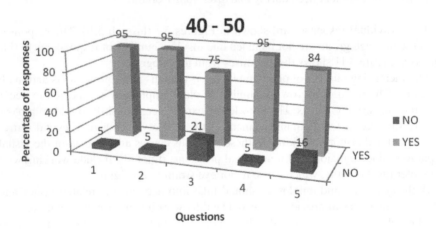

FIGURE 5.9   Percentage of responses from the 40- to 50-years-of-age category

of OHS is exclusively on young employees. The issue of older employees has so far been less prominent.

The legislation related to occupational health and safety defines ways to minimize risks. However, the legislation is not able to cover all the specifics of different work-places. It is thus the duty of management staff to identify what may be hazardous to employees at a workplace, to assess the frequency and potential consequences of the hazards, and eventually, adopt measures to reduce the risk.

The outcomes of the survey unequivocally show that eyesight deteriorates with growing age. The impaired eyesight and/or fatigue induced by eyesight-harming activities may be the cause of the occurrence of undesirable events, such as falls in the workplace. If employers keep hiring an even older labour force, they must be familiar with ways to modify the activities and environment to reduce the hazard (risk) level for these employees.

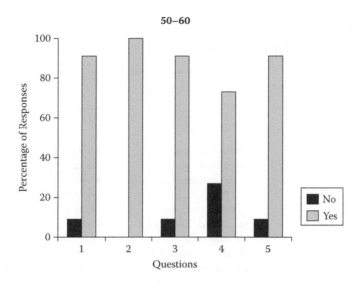

FIGURE 5.10 Percentage of responses from the 50- to 60-years-of-age category.

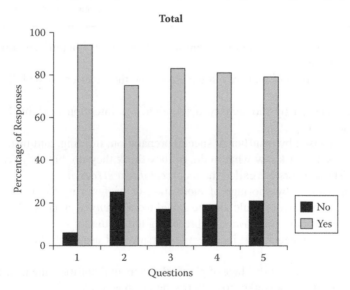

FIGURE 5.11 Total percentage of responses.

5.2.3 Do Changes Related to Aging Affect the Reliability of the Human Factor?

Changes in work abilities are closely connected with the reliability of the human factor within the Man–Machine–Environment system. The assessment of reliability is a very difficult process, since each person is an individual entity whose actions are affected by a variety of factors, for example, experience, skills, the ability to

TABLE 5.1

Human Errors of Older Employees

Changes Related to Aging	Work Environment Factors	Human Errors
Changes of neuromusculoskeletal function and functions related to movement	Job description/physical workload	Errors caused by shortage of physical and mental abilities
Changes in mental functions (depression, anxiety, fear of being laid off)	Stressful and dangerous working environment	Errors caused by momentary inattention
	Working environment layout	Errors caused by shortage of motivation or by ignoring the work procedures
Changes of functions of metabolic, digestive, and endocrine systems	Stressful and dangerous working environment	Errors caused by momentary inattention
	Working environment layout	
Changes of sensual functions	Stressful and dangerous working environment	Errors caused by momentary inattention
		Errors caused by shortage of physical and mental abilities

objectively judge one's potentials, physical dispositions, health conditions, and actual mental state.

The most significant kinds of human errors and their causes are as follows:

- Errors caused by momentary inattention; the intention is right, but performed incorrectly.
- Errors caused by insufficient special preparation, training, and instruction; workers do not know what to do, or they think they do, but they really do not. These errors are called the *incorrect intent errors*.
- Errors caused by shortage of motivation or by ignoring the work procedures. These errors could also be called *work offenses*, since workers making the error are fully aware of breaching the regulations.
- Errors by management; incorrect leadership and usage of plans, training, and skills.
- Errors caused by a shortage of physical and mental abilities; the insufficient prerequisites of a worker to perform the given activity.

The human error intensity will probably grow with age. The reason is a gradual decline of physical abilities. Examples of potential human errors related to older employees are shown in Table 5.1.

5.2.3.1   Accident Rate of Older Workers

Analyses of fatal industrial accidents (see Figure 5.12), serious industrial accidents with grievous bodily injuries (see Figure 5.13), and accidents with sickness absence

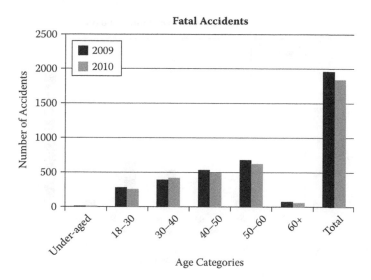

FIGURE 5.12 Fatal industrial accidents by age categories.

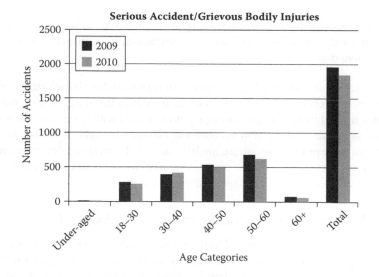

FIGURE 5.13 Serious industrial accidents by age categories.

of more than 42 days (see Figure 5.14) show the age structure of employees who suffered a serious industrial accident in Slovakia in 2009 and 2010.

The given analyses show that:

- The highest number of fatal accidents in both 2009 and 2010 occurred in the 50–60 age group.
- The second highest number occurred in the 40–50 age group.

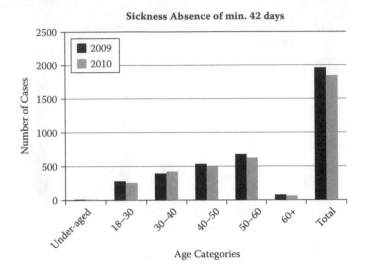

FIGURE 5.14 Industrial accidents with sickness absence of more than 42 days by age categories.

- Employees of the 40–50 and 50–60 age categories suffered the highest number of serious industrial accidents and accidents with sickness absence of more than 42 days.

The statistics also show a higher number of serious industrial accidents suffered by employees over 40 years of age. In order to reduce the frequency of industrial accidents, it is required to create a working environment with the emphasis on higher levels of safety and taking into account the changes brought about by the aging process. To achieve the appropriate modification of the working environment, it is necessary to be familiar with the changes in employee abilities and potentials and with the physical and mental changes that cause those.

5.2.4 STRATEGY AND APPROACHES RELATED TO OLDER LABOUR FORCE

'Older employees' may be included in the group of factors affecting the occurrence of emerging risks. The reasons for this are the changes of physiological qualities of human beings as a time function, and thus their impaired ability to respond to certain aspects of the working environment.

The following are among new risk factors:

- Factors, the existence of which has not yet been recorded, that is, these are created by new processes and technologies, new types of workplaces, and by organizational changes at work.
- Factors that have long-term effects (the given conditions have been known for some time), but only as a result of the latest scientific discoveries or upon changes in public opinion, are the factors considered as hazardous, or as hazard-causing dangers that are unacceptable and warrant correction.

Hiring older employees is among the new risk factors. This is supported by the fact that a human organism naturally gets older and the aging process causes irreversible physiological, and potentially mental, changes. The changes may be visible, though in a majority of instances these changes are gradual, and as such, are difficult to spot. The changes often lead to a decline of the worker's abilities and potential. These employees themselves normally do not realize or admit impairment of their abilities. It is therefore essential for management staff to estimate the need for modifications of the working environment, working hours, and responsibilities for the given group of employees.

For the management staff to perceive the changes and respond accordingly by implementing a variety of measures, they must possess the necessary (at least basic) knowledge of the changes typical in the aging process.

Imbalance between the physical working requirements and the employees' abilities during their active life occurs as a result of a decline in physical abilities and changes related to aging, which is not matched by the modifications of the job requirements. This creates grounds for occurrence of undesirable events and is a source of dangerous situations—hazards. The fact that the decline in physical abilities goes hand in hand with the aging process has been known for years, however, it is only today that the necessity to come up with solutions in this area has been given priority.

All this stems from a constant increase of the active age of employees as required by the European Union, from the efforts of individual countries to stabilize their pension systems, and from the need of organizations to compensate for the lack of a young labour force. The increase of the retirement age is causing the aging of the labour force; this very much affects the area of OHS in which the increasing age factor is included in the emerging risks from the point of view of their influence on a certain group of employees.

Identification and control of new and emerging risks is currently the subject of various scientific projects. The reason is the necessity to identify new dangers and hazards that organizations face prior to the occurrence of ensuing undesirable events.

The following are the questions to be asked concerning this matter:

- How will individual changes be manifested in the working environment and, in particular, industries in relation to performing working activities, including OHS?
- Which of these changes will be significant for the area of OHS?
- How should older employees be approached so that these changes are taken into account during working activities with the aim to avoid discrimination?
- How should older employees be motivated to remain employed?

For the field of OHS, the following is the most important question that requires a more profound analysis: What will be the effect of changes related to hiring older employees on the probability of an undesirable event occurrence and potential consequences as risk parameters?

It is obvious that the above-mentioned physiological changes will affect both parameters. The conditions of older public transport drivers may serve as an example. The factors influencing the occurrence of an undesirable event include, among others, the changes in mental and sensual functionality; in cardiovascular, haematological, immunological, and respiratory systems, in neuromusculoskeletal functions; and in functions related to movement.

Potential consequences of a public transport driver's failure are, in a majority of cases, fatal (see Table 5.2). Therefore, it is vital to possess detailed data on the physical conditions of older employees and on individual changes. The higher probability and consequence of an undesirable event occurrence with employees over 50 years of age must be included in the risk control processes in companies, with an extra effort put in creating a safe and healthy working environment for the group of older employees.

Particular goals related to an active aging process will not get stuck in political discussions; the goals will make a great impact in organizations that will have to focus their activities on the intents and goals to which the individual member states commit.

Motivating older employees to remain employed requires, most of all, improvement of working conditions and their adjustment to health conditions and needs of older employees. This requirement has been included in KOM (2010) 462 Final in relation to the year 2012 being declared European Active Ageing Year by the European Commission.

The individual countries will have to define their commitment and activities to improve the working environment of this group of employees. An improved working environment should be understood as an environment that is friendly to employees' needs and does not create any ground for the occurrence of industrial accidents and operational diseases.

The commitments represented by legislative changes at the international and national levels, labour force aging resulting in a shortage of active labour force, the shortage of an experienced and educated staff, financial costs connected to improving the qualifications of young employees, and the reduction of industrial accidents and occupational diseases are all factors that push organizations to realize the need for changes in favour of older employees. Therefore, these factors are labelled *push* factors.

The other factors are *pull* factors, which unlike push factors, do not create any pressure on organizations' managements and are a driving force in the process of achieving ever-improving key performance results. The pull factors include economic benefits to an organization (the higher productivity of older employees in a working environment favourable to their needs), improved competitiveness, image, motivating employees by creating a healthy working environment for all age groups, staff loyalty, and so on.

Provided that push factors create pressure on an organization or the organization creates extra space via pull factors, it is important for necessary systematic changes to be applied.

The whole process of creating a working environment that is favourable to the needs of older employees with respect to OHS inspired by the plan–do–check–act (PDCA) cycle is a simple and logical sequence of steps aimed at creating a healthy and safe working environment (see Figure 5.15).

One of the inseparable parts of each organizational change, and thus a basis for effective planning, is the analysis of the current state. For organizations to support

TABLE 5.2

Impact of Changes Associated with Aging on Probability and Consequence of an Undesirable Event

		Public Transport Driver	
Changes of Functions	Symptoms of Changes	Influence on the Probability of Undesirable Event Occurrence	Influence on Consequences
Mental functions	Difficult space orientation, delayed reactions, impaired short-term memory, loss of specific reflexes, …	Higher probability of an accident due to: • insufficient level of attention, delayed reactions to traffic, stress …	An accident with injuries to or death of the driver and/or other people in traffic
Sensual functions and pain	• impaired visual acuity • impaired long-distance vision • impaired ability to recognize colours • dry-eye syndrome • impaired auditory acuity • problematic localization of sound sources, pressure sensitivity …	Higher probability of an accident due to: • delayed recognition of items and people in traffic • eyesight fatigue, watering eyes • failing to recognize warning colour signs • failing to recognize warning signals • failing to identify the source of sound …	Accidents with injuries to or death of the driver and/or other people in traffic

Continued

TABLE 5.2 (*Continued*)
Impact of Changes Associated with Aging on Probability and Consequence of an Undesirable Event

Public Transport Driver

Changes of Functions	Symptoms of Changes	Influence on the Probability of Undesirable Event Occurrence	Influence on Consequences
Functions of cardiovascular, haematological, immunological, and respiratory systems	• increasing blood pressure • slow filling of heart chambers • significant increase of systolic pressure during work • impaired immunity against bacteria and viruses • increased minute ventilation and oxygen consumption during work • bad conditions for airways muscle functionality	Higher probability of an accident due to: • increased likelihood of heart attack at older age when exposed to stressful situations • chest pains • insufficient oxygenation • increased likelihood of sickness • absence at older age • increased likelihood of long-term treatment	Accidents with sickness absence, viral diseases
Neuromuscular-skeletal functions and functions related to movement	• arthrosis, osteoporosis, ligaments and tendons less durable • reduced body circumference • less strength in shoulders and fingers	Higher probability of an accident due to: • joint and muscle pain • impaired movement • decline in physical abilities • increased likelihood of fractures	Accidents with injuries to or death of the driver and/or other people in traffic Fractures Stretched ligaments and muscles Occupational diseases

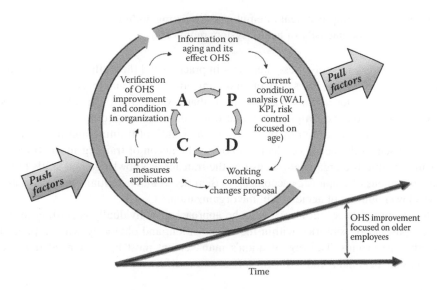

FIGURE 5.15 Push and pull factors.

changes for the sake of older employees, they must have at their disposal all the necessary information on the reasons and arguments in favour of the changes being carried out. The analysis, for this matter, should include information not only about the status quo within the organization, but also the information on demographic conditions in the country and on the aging process and related changes.

Failing to collect the above-mentioned data is likely to result in the proposed changes not having the required effect for the organization.

Another important step is getting familiar with the status quo within the organization itself. The analysis could make use of labour force profile analysis, the Work Ability Index (WAI) as elaborated by Prof. Ilmarinen and successfully used in Europe and other countries, key performance indicators (KPIs) focusing on an older labour force, and a Working Conditions and Control Questionnaire (WOCCQ) designed to assess employee stress levels.

Based on detailed information, it is possible to reconsider the need for realization of individual changes. In order to make the particular changes more specific, the human Failure Mode and Effects Analysis (FMEA) method could be used to analyze the causes of human errors, followed by a proposal for corrective measures related to organizational changes, changes in the ergonomic layout of a workplace, and so on. When designing changes in working activities, virtual reality (VR) may be used to test the given measures without putting employees in real danger.

The following steps could lead to creating a working environment that respects the needs of older employees:

- Improve the concept of tasks
- Improve work organization
- Improve the real working environment

- Support the improvement of employees' working abilities
- Enhance the intensity of health system monitoring

It is necessary to apply these measures in practice and their efficiency should be monitored in relation to the safety of an organization's older employees.

The condition for a functional approach that respects the older labour force is constant education and training aimed at this group with respect to the changes in mental functions, for example, by adjusting the length of training and the training process itself to the needs of older employees. Inclusion of training in the lifelong learning within the organization ensures the transfer of knowledge and skills to all the generations of employees according to their needs and potentials, and translates into a lower number of accidents in the organization.

Respective changes, however, must be approached individually according to the conditions and possibilities within the organization, and obviously with respect to specific age groups. The organization's management must be convinced that the changes are beneficial and will bear an economic effect.

5.2.5 Virtual Reality as Part of Risk Management Systems

The mutual relation between ergonomics and safety related to the aging labour force creates space for utilization of VR and application of the technology in designing solutions for maintaining older employees in employment.

Modelling a working environment using virtual reality helps design, in real time, a model reality that does not actually exist. This way, for the area of OHS, it is possible to analyze the existence or occurrence of hazards in the Man–Machine–Environment system at minimal cost and with no real hazard to health.

The virtual reality application could be considered a new and progressive tool to be used in the risk management system.

5.2.5.1 Virtual Reality

Virtual reality simulates a real or modelled environment that is possible to be visually perceived from the point of view of height, width, and depth, and as such, could provide a visual experience in real time, including sound, touch, and other forms of feedback.

The strategy of VR systems is to visually create a computer-simulated environment that is similar to the real world. Individual applications provide for displaying a working environment that emulates the real environment (see Figure 5.16).

Using other special equipment, such as stereoscopic goggles, video goggles, virtual walls, data gloves, and so on, enables us to create almost a real picture imported from the software-generated design. The individual VR applications differ in the input and output of particular data (see Figure 5.17).

To make the picture real and dynamic, the equipment must be able to display the individual frames in a fast sequence so that a user gets the impression of fluent movement. The higher the frequency of the computer monitor, the higher the picture quality. The human eye is able to perceive approximately 20 individual frames per

FIGURE 5.16   Designing a working position using virtual reality.

FIGURE 5.17   Simulation of environment using additional equipment.

second. A realistic impression is achieved at the level of 50 frames per second. For a 3-D picture, each frame must be different for each eye, or stereoscopic.

VR is utilized especially in industries that require a 3-D analysis and display of physical measurements.

VR is being used more frequently for various purposes, especially in the following areas:

- Education and training: individual or group training using virtual equipment and processes; training with products before their actual construction
- Medicine: virtual surgery (new surgery techniques) and virtual laboratories
- Entertainment: interactive 3-D games and 3-D theme parks
- Architecture: city visualizations, urban planning, interior design of 3-D buildings and equipment, and so on
- Production: product design and construction, maintenance, virtual prototypes, and ergonomics

The simulation displays 3-D objects or processes via computer technology. The visualization offers the user a complete concept of the product or process in question. The simulation process, with the addition of interaction with a user, gives a lifelike experience and helps them understand the data presented in a form of human eye–friendly, 3-D object representation.

By means of VR, it is possible to present the Man–Machine–Environment interaction, in which a man, the human factor, plays a significant role from the OHS point of view, without real danger to people or without causing any other loss.

5.2.5.2 Role of VR in Assessing Emerging Risks

In the process of risk and human reliability assessment, there are various methods to be used, such as Cognitive Reliability and Error Analysis Model (CREAM), A Technique for Human Error Analysis (ATHENA), and Technique for Human Error Rate Prediction (THERP).

Following are the drawbacks of these methods:

- The outcome of the assessment depends on the person conducting the assessment (competence, expertise, deduction …).
- The outcome of the assessment discloses merely potential failures for predefined activities, excluding accidental events, and so on.

Unlike these methods, virtual reality broadens the possibilities to identify potential human factor failures and negative event occurrences.

5.2.5.2.1 Importance of VR in the Process of Risk Assessment and Control

Virtual reality will have a significant impact on the area of risk assessment. Based on the available data, fifty-eight percent of all dangers identified in the field of engineering technologies are identifiable via VR. The real environment simulation has helped identify twenty-five percent of dangers and modelling a human being has contributed to detecting ten percent of all identified kinds of dangers.

For the VR application in the process of risk assessment (Figure 5.18) to produce desirable results, it is essential to implement all available information about the changes related to the aging of a human organism in the process of modelling a human being.

Decline in the size of the body (height and body fat distribution), decline in muscle mass followed by decline in physical strength, metabolism changes, changes of vision fields, impaired hearing sensitivity, and changes in joint flexibility will all have a significant impact. Using VR, it is possible to define activities that may cause problems to this group of employees and may be harmful to the employees or others.

The limited ability to work creates grounds for negative event occurrences. A VR-generated 3-D environment simulates real conditions and, as such, is able to emphasize the situations that may cause hazards, and this is how VR could contribute to minimizing risks of older employees' work.

As far as the older generation is concerned, VR is currently used mainly in the field of education and training, though it may also be used to analyze a user's behaviour in potentially dangerous situations without any real risk.

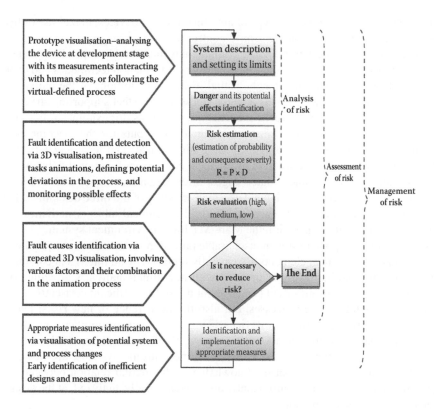

FIGURE 5.18 Applying VR in the process of risk assessment and management.

5.2.5.3 Partial Conclusion

Development of modern technologies is the ground for advances in a variety of scientific fields. Using the VR technology is easier, simpler, and, most of all, more effective in the processes of planning, designing, and producing safe machinery and workplaces. Application and development of this technology will be of significant importance in the area of risk management.

5.3 RISKS IN MECHATRONIC SYSTEMS

Control and information techniques are these days a part of the majority of machinery equipment and systems.

If there is a failure of electronic components, a car will not start, power plants will shut down, power supplies will be cut off, and it will be impossible to use transport and handling equipment within complex logistic systems. The electronic system in Daimler Benz's (GER) Maybach car consists of seventy-seven electronic control elements with 200 electronic contacts used to interconnect various internal systems of the car. Top-class models of BMW and Audi use a similar amount of electronics, and VW's Phaeton uses sixty electronic control elements. Modern intralogistic equipment, such as the travelling type of automated lifting devices, makes use of five dashboard computers and a single central control unit.

Functionality of modern intralogistic equipment is increasingly dependent on electronic components in relation to the functionality of information techniques and technologies. Requirements to monitor the actual technical conditions of the individual parts of a machine, or a machine as a whole (for example, steel supporting structure of a travelling crane), from the real-time machine responses (for example, robots and their control systems), and transmission of signals as inputs for the machine control units, are getting to be of ever-increasing importance. The term *mechatronics* defines, in particular, the extension of mechanical component functionality by means of integrating electronics, and in later stages, information techniques and technologies. Nowadays, the term is not defined as a mere synergy of the three disciplines, but also as the term defining brand new principles of the technical subjects' activities and approaches to their analysis within the Man–Machine–Environment system.

Mechatronic components are an inseparable part of modern technical equipment. Old-fashioned mechanical elements are partially replaced by an intelligent combination of software and hardware, or updated with new functionality. A growing portion of the IT and electronics used brings about not only significant cost-cutting and optimization factors in the processes, but also the occurrence of new risks! Faults and errors that result in modern devices being withdrawn by mother companies for system repairs and modifications go to prove the fact that the mutual relations between all the structures of modern devices, mechatronic systems, are not paid enough attention. It follows that the knowledge of the mutual relation between the individual structures are not consequently and systematically analyzed, at the design stage or at the stage of operation.

Mechatronics as a discipline of engineering is built on the fact that, as a result of mutual actions of various technologies, new functions come into existence. Based on the integration of mechanics, IT, and electronics, it is necessary for information techniques, elements of control and regulation techniques, material engineering, and linking building techniques to be added to the mechatronics systems.

Each mechatronic system consists of the basic mechanic structure, sensors, actors, and information processing. These parts communicate via the information flows that allow for control and regulation of the systems and create conditions for implementing new functions. Through sensors, the data defining the actual conditions of the device, process, or environment are monitored. Based on the assessment and processing of these signals in the central unit, commands are issued to the actors who carry out a change in the system so the system is back to its optimal operating conditions. This is how the conditions for intelligent and self-optimized systems are created (artificial intelligence). While at the level of the basic mechatronic structures the response to parameter changes is carried out via a simple regulation circuit, the interconnecting and networking of the individual mechatronic components significantly increase the system complexity. Therefore, the mechatronic system-level assignment is based on the level and quality of the information flows.

The solution for product development based on user requirements is an exact definition of the system being designed. This gets rather complicated with mechatronic systems, since with these interconnected structures there are a number of mutual influences. Furthermore, various engineering disciplines overlap from the point of view of terminology, functionality, and their construction. All these elements are characteristic for their different approaches to product development and, thus, to the application of risk assessment methods.

The highest level of network-interconnected mechatronic systems defines complex closed structures, for example, material flow centres for goods distribution, warehouses (logistics centres), or sorting devices. These are made of the structures that could operate independent of each other and utilize clearly defined functional interfaces, for example, transport devices (a roller conveyor, a conveyor belt, or a pan conveyor or sorting device modules). The autonomous mechatronic systems (AMSs) use the information processing systems, control devices, and fault diagnostic systems. AMS consists of mechatronic function modules (MFMs) as a basic structure of the mechatronic systems. The limits of a single-function module depend on the respective functions. On the basis of the unequivocal definition, the individual elements are assigned to multiple MFMs.

Provided that the software takes over the control functions of mechatronic systems to a large extent, the failure and risk analyses must be based on different principles than with the standard mechanical systems in use, for example, steel supporting structures. The mechatronic system could include several intermediate states that may cause the system to return to the initial position. It is important to know that during the technological lifespan of the mechatronic system, there may appear multiple time-independent failure events.

As a result of growing software complexity within mechatronic systems, the risk analyses applied to the mechanical elements also include, to a large extent, analyses used to define failures/risks of software packages. When using a phase model in the software development process, all relevant steps are implemented—the data processing concept, development, individual testing, overall testing, and the handover. Each stage defines the activity that needs to be used and how it is supposed to be documented. The analysis of all the intermediate states of the individual stages will show how the defined data have been followed as well as compliance with the requirements for the software system.

5.3.1 Application of Risk Management Methods in Mechatronic Systems

The aim of all the measures within risk management must be the elimination or reduction of risk of accident and/or shutdown with injury during the expected lifespan of a machine or machinery, including its assembly and disassembly, and during the unforeseeable events that may occur during operation.

It is obvious that the most important requirements for safe machine operation as a mechatronic system are placed at its design stage, at which all the risk analyses must be carried out and all the available technical measures must be implemented based on the outcomes. The goal herein must be the elimination or minimization of risk. The requirements also affect the selection of effective methods for risk estimation for all the parts of a machine or a complex machinery system, including its control system, at all the stages of a machine's technological lifespan.

Risk management includes:

- Defining the machine as a mechatronic system
- Identifying hazards to individual functional structures of the machine
- Estimating risk in the individual structures
- Assessing risk for the whole machine
- Calculating measures for risk minimization

The data necessary for risk assessment include:

- Definition of the functional structures of the machine
- Information on the individual stages of a machine's technological lifespan
- Drawings and other documentation depicting the machine functionality
- Detailed information on the history of failures and accident frequency and consequences (for example, accident records, information on faults and harmful processes as a result of machine operation)

As far as the application of risk management methods is concerned, machines must be divided into individual structures in order to exactly define mechatronic aspects with clearly defined functions within the overall function of a complex device. This way, the individual types of risks can be defined and, consequently, effective approaches to risk reduction can be selected as part of user manuals, or as a part of the process of creating the operating conditions within the logistic system.

Based on the fact that a mechatronic system is typically a synergic integration of mechanical and electronic elements, supplemented by intelligent control systems using IT techniques and technology, the potential risk analysis must be carried out within all three areas. The term *synergy* is of significant importance; the mutual interconnection means that failure (malfunction) of any of the areas causes a shutdown or failure of the whole system.

A mechatronic system is schematically defined in Figure 5.19 and is realized as follows:

1. Traditional machines are attached to an electronic control system with IT techniques or intralogistic tools, for example, lifting devices, a combination of conveyors, computer numerical control (CNC) machine tools, robots, manipulators, and so on.
2. Some mechanical functions of the machine are replaced with electronic controls, for example, replacing mechanical limiters of the lifting device loading capacity with the electronic ones in compliance with IT, and replacing mechanical gearboxes with electronic systems–controlled ones.
3. Mechanical control functions are replaced with electronic ones that use IT techniques.

The system risks of a mechatronic system are characterized by the following four basic groups with a different risk occurrence mechanism in each area:

1. **Mechanical risks:** Failure of mechanical components of the system

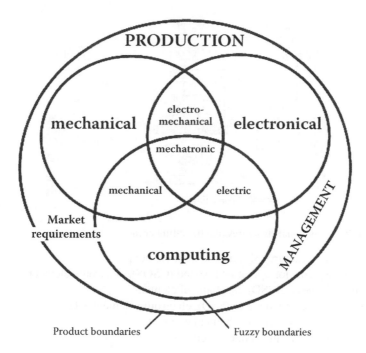

FIGURE 5.19 Principle of the mechatronic system.

2. **Electric and electronic risks:** Power shortage, malfunction of electronic components, malfunctions of testing and control systems
3. **Risks within IT systems:** Occurrence of a random logistic fault, intentional or unintentional software modification, and incorrect input information
4. **Risks as a result of synergy failure:** An important aspect of mechatronic systems risk assessment is a mutual connection of the individual components. Even a minor, hard-to-define failure of a part of the system could cause, following incorrect assessment, a shutdown of the whole system followed by accidents or injuries. Malfunction of a mechanical element that cannot be detected, for example, due to the absence or malfunction of a sensor, may also cause an accident or injury.

5.3.2 Lifting Machines as Mechatronic Systems

A travelling crane, as a typical example of a lifting machine in both traditional and automated versions, is among the devices that represent a mechatronic system within the logistic systems structure. For the purpose of risk assessment in particular areas, a travelling crane as a mechatronic system could be divided into functional structures, as shown in Figure 5.20.

The design of the travelling crane as a mechatronic system presumes several independent functional structures that are unified into a single machinery unit to provide for material handling as a final activity of the lifting machine. The structures include:

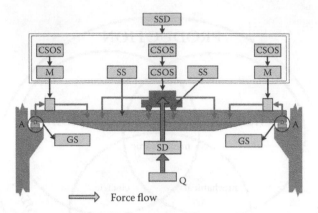

FIGURE 5.20   Functional structures of a travelling crane.

- Supporting structure (SS), mechanical element
- Control systems for operational speed (CSOS), electronic element
- Suspension devices (SD), mechanical element
- Safety and security devices (SSD), electronic element + IT
- Guiding systems (GS), mechanical element
- Mechanisms (M), mechanical element

The requirements for the individual structures, as governed by valid legislation, are vast and various, for example, control systems redundancy (electronic principle), technical diagnostics for drives and steel-supporting structures (mechanical and electronic principle), limit switches for limiting movement (electronic principle).

A travelling crane is typical for the following types of dangers that occur after the inclusion of the crane in a logistic system:

- Oscillation of the load attached to the suspension device, diagonal pull
- Handling activities on the suspension device
- Faulty ropes
- Falling off the crane
- Collision of two cranes
- Faults on the steel structure, crane overloaded

Apart from 'falling off the crane' and 'handling activities on the suspension device,' all other dangers can be characterized by the mechatronic principle.

The oscillation of the load attached to the suspension device, diagonal pull (SD, M): The oscillation occurs as a result of the movement of the crane or crab, that is, the electronic control systems applications. Reducing this kind of risk is done by activating the electronic control system of the moveable mechanisms of the crane bridge and crab (M) (see Figure 5.20). To prevent the diagonal pull, it is possible for the electronic equipment to be implemented in the lifting mechanism control system (Figure 5.21), where interconnection of the mechanical and electronic elements occurs during the resulting movement.

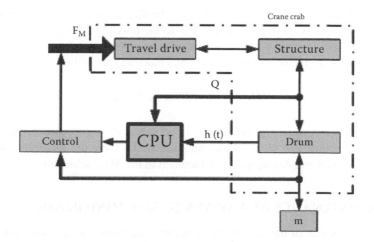

FIGURE 5.21 Principle of the electronic loading capacity limiter.

Faulty ropes (SD + faulty steel structure, M): The major cause of this danger is the crane suspension device being overloaded, as the resulting type of danger stems from mechanical failure of the steel construction. A lifting capacity limiter has been used as an effective means of monitoring the travelling crane's overload; the limiter also includes a device for defining the loading collectives, that is, the sensors to monitor the load weight on the suspension device, and for the crab and lifting device's bridge positions. The principle of this device is built on electronic elements, while in the process of dynamic overload identification, IT techniques are also used for programming the comparative value in the electronic loading capacity limiter (Figure 5.21). The malfunction of this device results in losing control of the actual load on the bridge steel structure and on the crab suspension device, normally a steel rope, and of the residual lifespan of the solid functional structure, the steel structure. This is followed by an incorrect definition of the probability of the steel structure failure and, thus, by increased risk.

A collision of two cranes: This type of risk is directly related to the functionality of the device that monitors the proximity of two cranes. The failure of the device causes increased probability of two cranes colliding. The collision may result not only in the functional failure of the lifting machine system, but also in injuries caused by potential derailing of the crane from the guiding structure (GS).

5.3.2.1 Application of FMEA Method for Lifting Machines as Mechatronic Systems

For the purpose of mechatronic systems risk assessment, it is possible to utilize a modified version of FMEA (see also Chapter 4.3). Following consultation with experts in the field of specific application of a lifting machine, the particular functional structures are assigned points in the range 1–10. For mechatronic systems, it is possible to apply the ME (mechanics and electronics) structure identification coefficient, also within a 1–10 range; the coefficient is directly dependent on a number of electronic components. If an element of the mechanical part of the machine is dealt with, ME =

1 is applied, that is, it is a gradual, easily identifiable hazard. With a growing portion of electronic elements, ME rises all the way to the maximum level of 10, accidental failure (from 0 to 1) of electronic elements or inappropriate software.

MR/P risk is then calculated for individual hazard types as follows:

$$MR/P = VZ \times PV \times PO \times ME \tag{5.1}$$

where MR/P = risk degree/priority, VZ = importance, PV = occurrence, PO = probability of detection, ME = mechanics and electronics structure identification coefficient, with electronic features and IT included in the ME coefficient.

5.4 MAINTENANCE AS A MEANS OF RISK MINIMIZATION

On the basis of one of the definitions of maintenance that says that maintenance is a set of activities to ensure safe operation, technical conditions, preparedness, and economical operation of elementary funds, the tasks for developers, designers, and users of all kinds of elementary funds are defined. This means that attention must be paid to keeping production technology elements and a final product safe and reliable during all the stages of its technological lifespan, including maintenance and repairs. It follows from the fact that the complex Man–Machine–Environment system and mutual influence of the system's components must be a subject of safety analyses. Risk reduction strategy must be implied as early as the projection and design stage. At this stage, conditions for maintenance activities on the technical equipment must be taken into account so that potential risks are eliminated or reduced while the activities are performed. Operational instructions for a machine must include the mention of residual risks during maintenance and repairs. For these activities to be effective, the cooperation between manufacturers and users is essential.

Basic principles of today's machines and machinery systems design and production processes include the following:

- Technical failures, regardless of when they occur, must not cause subsequent risks. A device, machine, or complex equipment must be designed so that failure does not spread in case there are several failures occurring simultaneously.
- Technical systems must not accept information leading to the occurrence of a dangerous operational condition.
- Technical systems must be adapted to the operational staff's qualification levels.
- User's manual must include the mention of all residual risks and all hazards that may occur during operation, including the maintenance stages.
- Material wear characteristics or aging of elements crucial to the safe operation of a machine must be defined precisely. These are the bases for assessing the machine upon its technical conditions, and are also the prerequisites for applying the maintenance and repairs strategy.
- Design of the maintenance activities must correspond to the user's capacity.

The role of maintenance has been dramatically modified as a result of the increasing significance of machine safety. The occurrence of a fault may cause a machine to be shut down, but it also naturally causes particular types of hazards in accordance with STN EN ISO 121001-2. Regarding this, Directive No. 42/2006/EC* in Annex 1, Section 1.6 also defines requirements in the area of maintenance and repairs.

Nowadays, the term *hazard* or *degree of hazard*, that is, *risk* is widely used in the field of maintenance and repairs. The definition of the term, in compliance with legislation that is valid in almost all corners of the world, makes it possible to monitor both the frequency of failures and their consequences, which is a radical change in the strategy of maintenance planning.

Eventually, some of the symptoms of insufficient machinery maintenance have a significant effect on the occurrence of injuries or semi-injuries. This includes:

- Material aging processes, for example, steel, plastics, sealing materials, and fatigue of material during dynamic operational loading
- Corrosion, surface corrosion, and corrosion on grain edges
- Electrolytic effects
- Wear as a result of friction
- Aging of lubrication agents

In relation to activities within the maintenance area, risk assessment systems are applied to:

- Risks occurring during maintenance activities
- Risks as a part of maintenance strategy
- Risks as a result of insufficient maintenance

Quality and safety of production technologies and final products are a function of time, operating time, and intensity of external factors, that is, the safety of a final product is a multiparametric function.

On the basis of this assumption, it is essential to monitor and inspect the machine/equipment safety and maintain stable conditions for risk reduction. Safety management is not a static issue—it is a continual task. After the guarantee period expires at the latest, it is obvious that the machine user is legally and morally responsible for activities within risk management (Directive 89/391/EC). The user's scope of activities, naturally and to a large extent, includes maintenance. A technical equipment operator must be familiar with the methodology and tools of risk analysis and must be provided with a management structure that enables machines and equipment to be operated in a safe way.

Based on the aforementioned reasons, it is essential for all maintenance activities to be applied so that during machine or complex machinery's technological lifespan, the conditions for its safe operation are secured, that is, constant application of active measures for technical and human risks minimization.

Since modern maintenance methods are typical for their interdisciplinary nature, they include modern methods of technical diagnostics and mathematical statistics

---

* European Commission.

are applied. It is necessary to respect the human factor and with a lifetime education process, elaborate programs that will reflect modern trends in maintenance technologies. The realization must be carried out by experts from universities, educational institutions, and research institutes, and this must be supplemented with presentations of findings and experience of experts coming from companies and firms that have had experience with implementing modern maintenance methods.

Outcomes of every single risk analysis of machinery today must also include:

- An obligation to maintain the machine or equipment in its idle state.
- The design of maintenance organization and strategy.
- The organization of spare parts supply must be in compliance with Directive No. 42/2006/EC.
- The operational staff's qualification requirements; an insufficiently trained staff represents one of the biggest risks.
- The proposal for operational staff's required qualifications.

It must be realized that an insufficient maintenance organization leads to incorrect maintenance activities/operations and this results in the inability to achieve optimal maintenance goals.

The fact that the technical systems' safety is a function of time means that their safety needs to be constantly analyzed and assessed. Activities of this kind include maintenance operations. Failure to sufficiently execute operations may result in an emerging technical fault causing an accident or injury.

Technical system failures may occur:

- As a function of time, for example, aging, corrosion, chemical processes
- As a function of operating conditions, for example, wear, fatigue of materials
- As a function of external factors, for example, emissions, electromagnetic fields, vibrations, dust

Assuming that the technical system must be safe in relation to operational staff or to third parties at every stage of its technological lifespan and during all the operating states, that is, risks must be minimized, the following must apply:

$$R_{sk} < R_{akc} \qquad (5.2)$$

where R_{sk} stands for actual, that is, real risk; and R_{akc} stands for acceptable, that is, allowed risk.

Risk minimization strategy must be applied as early as the projection and design stage. At this stage, the conditions for maintenance activities on given equipment must be taken into consideration in order for potential risks to be eliminated or minimized when the activities are performed. Operational instructions for a particular machine must include the residual risks during maintenance and repairs. For these activities to be effective, cooperation between manufacturers and the user is essential.

The golden rule says:

In order to prevent failure, it is essential for potential risks to be identified early, and for respective protective measures to be activated in a form that interrupts the causal relation of failure/injury occurrence.

For the maintenance activities to be performed effectively, it is necessary to be familiar with the actual technical conditions of the equipment. At the equipment's design stage, it is essential to:

- Define all significant risks (hazards) (e.g. in accordance with legal enactments, such as EN ISO 12100 Parts 1 and 2, mechanical, chemical, electrical, informational, during measurements).
- Determine, for each risk (hazard), the acceptable values and procedures for their identification (measuring) taking into account a minimal time delay between the technical condition identification and the activation of safety or protective measures on the machine.
- Design the protective measures to minimize risks.

During equipment operation it is, for the purpose of determining its actual technical condition, necessary to gather data depicting the operating conditions by means of:

- Collection of information based on the predetermined plan of inspections without the support of computing systems. The drawback of this procedure is the fixed interval of the inspections.
- Collection of information using a computer-aided system either continuously or at certain intervals. The drawback of this procedure is often insufficient software supply.

Note: Wear, aging, fatigue, and corrosion are processes, the spreading of which may, on many occasions, be described by mathematical models.

In case of the stochastic progress of damage, typical for electronic and microelectronic elements, the above-mentioned statements do not apply. Failure characterized by stochastic progress is described using statistical methods only. Accidental failure of this kind rules out the use of the above-mentioned strategies. In these cases, it is not useful to look for means to provide for safe operations in the form of high-quality maintenance, but in another form, for example, redundant connections.

Safe operation of technical equipment or machinery is not a constant quantity. During operation, changes occur as a result of the following parameters:

- Operating time (aging, corrosion, etc.)
- Operating conditions (wear, external factors, fatigue, etc.)
- Probability of failure occurrence (stochastic character, deterministic character, etc.)
- Transcendental states (electromagnetic effects, etc.)

For the above-mentioned reasons, it is essential for all the maintenance activities to be applied so that during the machine or complex machinery's technological lifespan, the conditions for its safe operation are secured, that is, constant application of active measures for technical and human risk minimization.

After each modification of the equipment, or after each medium-sized repair or overhaul, risk analysis is necessary to determine whether the equipment is safe and that risks during the ensuing operation are minimized. The following types of hazards are to be identified:

- Ergonomic (e.g. incorrect body position when operating the equipment)
- Long-term hazards (e.g. noise, vibration, radiation)
- Mechanical (e.g. being stuck, crushed, pierced)
- Electric (e.g. voltage, arc light, electromagnetic fields)

For hazard identification, it is advisable to use questionnaires. In the case of more complex and complicated equipment, experts in the field of machine safety must possess more experience and the methods used must be of higher complexity. Modern technical risk analyses are based on the collection of a vast amount of data that depict the technical conditions of the equipment. The pieces of information are further processed by computers, the central units, and are then compared with the acceptable values saved on the hard disk. The measures are implemented either automatically via the feedback connections of the operating elements, or are performed by the operational staff based upon the respective data. Based on the collected data, the appropriate maintenance is carried out and its strategy is designed to ensure safe equipment operation with respect to other additional goals, such as preparedness, reliability, space reduction, and so on.

5.4.1 Technical Diagnostics as an Effective Tool for Risk Minimization

Technical diagnostics is a set of activities performed to determine the technical conditions of technologies or their components. The aim of the diagnostics is the evaluation of the actual condition of objects based on the objective assessment of symptoms that are identified by means of measuring equipment, identification of degrading factors that could lead to equipment failure causing danger to the operational staff or other people outside the operating system of the technological subject.

Technical diagnostics is a tool for technical system risk minimization, and by using its methods, it is possible to recognize hazardous states as risk potentials.

Safety and reliability of a machine or equipment must be monitored and inspected systematically, and conditions must be created for minimizing negative events in a form that interrupts the causal relationship of the occurrence of an accident, failure, and injury, that is, minimizing risks. A technical equipment operator must be familiar with the methodology and tools for performing these activities and must be provided with an appropriate management structure in order for machines and equipment to be operated in a safe way. One of the effective methods of performing these activities is a set of activities within maintenance technologies including, in particular, technical diagnostics.

For the aforementioned reasons, it is essential for all the maintenance activities to be applied so that during the machine or complex machinery's technological life-span, the conditions for its safe operation are secured, that is, application of active measures for technical and human risks minimization is constant.

The risk minimization process calls for the application of procedures that are designed to identify the potential of failure occurrence at the hazard stage. Modern effective maintenance methods are typical for their interdisciplinary character and, as such, include to a large extent the modern methods of technical diagnostics.

Technical diagnostics includes in its name the term *diagnostics*, which stands for an activity that assesses the actual conditions of an object. Attention within these activities is focused on technical objects, that is, machinery systems, complex technologies, as well as individual machines.

Experts in the area of technical diagnostics focus their attention mainly on activities related to prevention of failures, accidents, and injuries as the most effective activities from the point of view of human and technical safety, and economy. The tools for creating conditions for minimizing risks, that is, reducing the frequency of failures, accidents, and injuries, include the application of a technical diagnostics method in a variety of industries as the most effective means of prevention activities.

Being familiar with the real conditions of the equipment enables a user to compare the equipment's actual operating conditions with the ones required by the manufacturer. The emphasis is on the ability to implement all the measures in a timely manner, and if possible, early enough for the equipment to be able to perform activities expected by the user. Failure of technical subjects may result in danger for all operators as well as other people within the environment affected by the failure (the principles of occupational health and safety are not adhered to).

When in operation, for the purpose of determining the equipment's actual technical conditions, it is essential to gather data depicting the operating conditions, that is, applying the technical diagnostics methods along with the latest methods of experimental measuring. It is also vital to verify the measured data as well as the data processing methodology to eliminate the risk of the data relevance to the monitored subject.

5.5   ACOUSTIC RISK MANAGEMENT

Today, reducing noise levels, and thus reducing oscillation levels, is generally considered one of the important goals in developed countries It is estimated that one-third of employees in Europe (over 60 million) are exposed to high noise levels for more than one quarter of their work time. The result of the illegal exposure of people to high noise levels is an irreversible process of auditory system disorder or extra-auditory effects, such as damaged centres in the central nervous system, loss of concentration, damage to eyesight, and so on, which eventually may lead to the decline of concentration when performing working activities and to subsequent injuries.

Activities in the area of acoustic risk prevention should mainly be focused on primary reduction of noise and oscillation, that is, on direct elimination of the sources of the significant vibroacoustic energy of the equipment, by means of, for example, design changes, the appropriate selection of materials, technological changes (e.g. replacing roller bearings with friction bearings of negligible noise levels, re-centring or rebalancing of rotating engine parts, and flexible connections of individual machine parts). Secondary (subsequent or additional) reduction of the vibroacoustic energy is less effective and more costly. It serves to absorb a certain amount of the energy. The secondary noise-cancelling measures include, for example, sound-proof shields and noise absorbers, sound-proof dividing walls and screens, noise-absorbing materials on the walls, and so on. Tertiary measures, as the last resource of noise-cancelling means, are normally implemented by the user of the equipment, for example, by using personal protective equipment (PPE). The most effective results are, however, achieved with the primary reduction of the vibroacoustic energy that is in the initial stages of the causal relationship of failure/accident occurrence. Vibroacoustic energy reduction methods could be applied right at the source in the process of designing, developing, and constructing new machines, machinery, technologies, and means of transport with the use of the latest scientific discoveries and knowledge.

It is man who most contributes to the level of vibroacoustic energy of a machine. A person is the one who determines the accuracy of individual parts of the machine, their balance, treatment, accuracy of technological assembly, selection of appropriate materials, absorbing fillers, appropriate shapes of piping for non-stationary flow, selection of a suitable technological action, and so on.

5.5.1 ACOUSTIC RISK MANAGEMENT ALGORITHM

Effective risk minimization is achieved only if an issue is dealt with systematically. When forming a risk reduction strategy and implying measures at new and existing workplaces, the following steps are to be taken:

1. Setting goals and criteria
2. Identification and assessment of noise sources:
    * Emissions at workplaces
    * Effect of external noise sources on immissions at workplaces
    * People's exposure
    * Noise source emissions to set the order of their importance
3. Evaluating noise reduction measures, such as:
    * Machine noise reduction
        - Reduction of noise transfer at a workplace
        - Noise reduction at workplaces
4. Designing the noise reduction program
5. Implementing appropriate measures for noise reduction
6. Evaluating the achieved level of noise reduction

When designing the protective measures it is advisable to follow the algorithm structure for noise reduction shown in Figure 5.22.

Following are the options for determining the integrated value of acceptable risk:

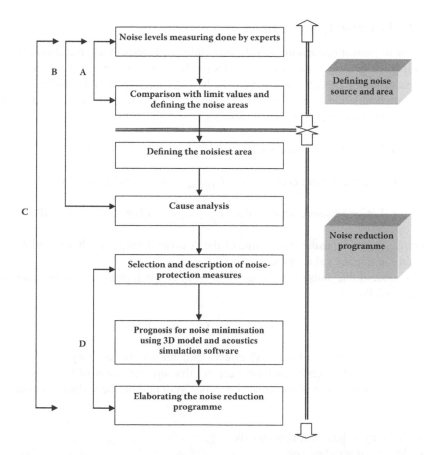

FIGURE 5.22 Acoustic risk management algorithm: A = risk analysis, B = risk assessment, C = risk management, D = risk control.

- The acceptability of the vibroacoustic environment could be considered as a tolerable degree of unfavourable conditions of simultaneous effects of noise and mechanical oscillation in working environment.
- Vibroacoustic acceptability of the environment is assessed by the subjective disturbance criterion, by the interference with a human activity and productivity or with occupational health and safety, and the combination of them.

Another option to assess acoustic risk is the application of the following formula:

$$R = P \times C, R = f(E_x, L_{eg}, C_{hearing}, C_{out\,of\,hearing}) \tag{5.3}$$

where R stands for acoustic risk, $P = f(E_x, L_{eg})$ stands for the probability of a person being exposed to noise; it is a function of these parameters: exceeding legal limits L_{eg} (dB) and time exposure E_x (hours); and $C = f(C_{hearing}, C_{out\,of\,hearing})$ stands for the consequence of acoustic stress effect on people's auditory organ $C_{hearing}$ and extra-auditory effects $C_{out\,of\,hearing}$, which are more difficult to identify.

5.5.2 EXCEEDING L_{EG} LEGAL LIMITS

In order to protect occupation health, especially with respect to protection against audible noise-causing reductions in staff's productivity, the limit and action values of noise exposure are defined, for example, by EU Directive No. 2003/10/EC, on the minimum health and safety requirements with regard to workers' exposure to risks caused by physical agents (noise), as follows:

a. Exposure limit values $L_{AEX,8h,L}$ = 87 dB and L_{CPk} = 140 dB

b. Upper exposure action values $L_{AEX,8h,a}$ = 85 dB and L_{CPk} = 137 dB

c. Lower exposure action values $L_{AEX,8h,a}$ = 80 dB and L_{CPk} = 135 dB

where $L_{AEX,8h,L}$ = exposure limit value of the A-noise during eight hours and L_{CPk} = exposure limit peak value of the C-noise.

The undesirable noise effects as risks affecting a human organism could be defined as follows:

$$D = f(C_{hearing}, C_{out\,of\,hearing})\qquad(5.4)$$

where $C_{hearing}$ = specific (auditory) effects that impair the auditory analyzer and $C_{out\,of\,hearing}$ = system (extra-auditory) effects that are the result of the changes in functions of other parts of the central nervous system than the auditory apparatus, and include:

• Inability to hear warning signals
• Increased accident rate
• Increased fault rate in a production process

Acoustic risk could be given the following values:

Z = Negligible acoustic risk; no measures necessary to be applied, but it is essential for the risk levels to be regularly inspected for changes.
A = Acceptable acoustic risk; acceptable under certain conditions (preventive measures of technical or organizational nature).
N = Unacceptable acoustic risk; immediate corrective action required in the form of effective measures to be applied.

The most common graphical representation of acoustic risk assessment is a matrix as seen in Figure 5.23.

5.5.3 NOISE REDUCTION STRATEGY

The goals set for the acoustic risk minimization process must be based on the fact that noise must be reduced to the lowest levels possible (allowed). These values could

P \ C	none	temporary	rare occurrence	remaining
L_{eg} not exceeded $E_x < 8$ hours	Z	Z	A	N
L_{eg} not exceeded $E_x > 8$ hours	Z	A	A	N
L_{eg} exceeded $E_x < 8$ hours	Z	A	A	N
L_{eg} exceeded $E_x > 8$ hours	Z	A	N	N

FIGURE 5.23 Acoustic risk matrix.

be represented by noise emission levels or by noise exposure levels. Normally, the values with the A parameter for noise emission or noise exposure are taken into account and must not exceed the legal levels as defined by Directive No.10/2003/EC.

Noise reduction may be realized by means of a variety of technical measures. The measures include source noise reduction (e.g. noise generated by machines, working processes and procedures), noise reduction by noise absorption (e.g. by using shields, screens, absorbing panelling), noise reduction on designated locations (e.g. soundproof cabins, personal ear-protective equipment).

The acoustic risk reduction measures may significantly modify the Man–Machine–Environment system. Therefore, it is advisable for the parties involved to make use of each proposal for measures and take an active part in their preparation process. This especially concerns people who are active in the fields of management, planning, purchasing, occupational health and safety, maintenance, technology, and production, as well as technical staff, union members, and workers themselves. On many occasions, it is necessary to involve external parties, such as hygiene and occupational health and safety inspection authorities, experts in acoustics, ergonomics, and so on. This kind of cooperation between company representatives and external parties will ensure that, when selecting the noise reduction measures, all the specific aspects related to the given project will be taken into consideration.

Successful noise reduction depends on active and committed involvement of the company's management.

Priority must be given to the prevention defined by legal enactments, such as Directive No. 42/2006/EC, on machinery. A machine in compliance with the directive must be designed and constructed so that risks caused by airborne noise emissions are reduced to the lowest levels possible with respect to scientific knowledge and technology as well as availability of means for noise reduction, especially at the source of the noise. The noise emission levels may be assessed using comparative data on noise emissions for similar machine equipment.

The user's manual, according to Directive No. 42/2006/EC on machinery, must include:

- The installation and assembly instructions, with the aim to reduce noise and vibrations.

- The information on airborne noise emissions including:
  - A noise emission level at a workplace as measured by weighted filter A, if the level exceeds 70 dB(A); if the level is below 70 dB(A), this must also be stated.
  - A maximum momentary noise level at a workplace as measured by weighted filter C, if the level exceeds 63 pa (130 dB with the reference point of 20 µPa)
  - A level of machinery acoustic power as measured by weighted filter A (normally weighted filter A is used, as opposed to weighted filter C; they follow the loudness curve for 40 and 100 phones, respectively; the curve of weighted filter A is closest to the curve of harmful noise determining the negative noise effects on the auditory system), if the noise levels at a workplace, determined by weighted filter A, exceed 80 dB(A).

The data must be measured for particular machinery or determined on the basis of the measurements coming from technically comparable machinery.

Regarding oversized machinery, instead of the level of acoustic power, as determined by weighted filter A, it is possible to state the noise level measured by weighted filter A at designated locations around the machinery.

Provided that harmonized standards are not applied, the noise levels must be measured using methods most appropriate for the machinery. Every time noise emission data are stated, the uncertainty of the measurements must be specified. The description of the operating conditions for the machinery during the measurements, as well as the description of the measuring methods, must also be made available.

If the workplace is not defined or is impossible to define, the noise levels determined by weighted filter A must be measured at a distance of 1 meter from the machinery surface and at the height of 1.6 meters from the floor or the platform; the location and maximum noise level must be stated.

5.5.3.1　Acoustic Study at Laboratory

Description of workplaces, working process, and activities:

The section of the laboratory where the measurements took place houses machine tools and devices (grinding machine, table drilling machine, lathe, and milling cutter). The whole area is shown in detail in the hall ground plan (Figure 5.24). The machine types located in the analyzed hall are described in Figure 5.25.

The real data measured are included in Table 5.3.

For assessing acoustic risk, the following could be used:

$$R = f(E_x, L_{eg}, C_{hearing}, C_{hearing}/C_{out\,of\,hearing}) \tag{5.4}$$

where L_{eg} = legal requirements fulfilled with no exposure limit values exceeded, apart from the case of the grinder and two ventilators working simultaneously; E_x = operational exposure for less than eight hours (training process and laboratory measurements); $C_{hearing}$ = no ear-related problems recorded; and $C_{out\,of\,hearing}$ = when working with the machinery, verbal communication is disrupted.

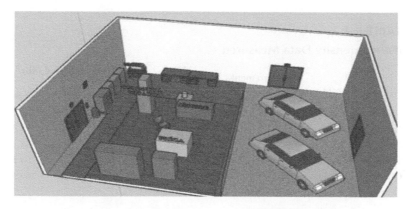

FIGURE 5.24   Hall ground plan.

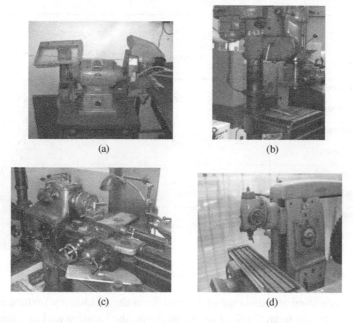

FIGURE 5.25   Types of machine tools located in the hall: (a) grinder, (b) driller, (c) lathe, (d) milling cutter.

It follows from the above that acoustic risk at the laboratory is negligible for the individual items of machinery, since the respective parameters have not been exceeded, and thus no measures are necessary to be applied based on Figure 5.26. Only when simultaneous action of the grinding machine and ventilators occurs is the exposure limit value exceeded, so the risk is increased to the acceptable risk level. In order to ensure safe work, it is necessary to use PPE only.

In compliance with Directive No. 42/2006/EC, the acoustic prediction concerning the intended position of machinery is important, or in case workplaces are not defined or impossible to be defined, the noise levels determined by weighted filter

TABLE 5.3

Noise Intensity Data Measured

Post	Measurement Time(s)	Status of Device	Data Measured (dB)		
			$L_{Aeq,8h,L}$	L_{Amax}	L_{Crpeak}
Grinder, Figure 5.25a	15	Grinder in action	85	89	102.3
Driller, Figure 5.25b	15	Driller is on	65.4	76.5	92.3
Lathe, Figure 5.25c	15	Lathe is on	76.5	80.1	91.7
Cutter, Figure 5.25d	15	Cutter is on	73.2	82.6	94.5
Grinder + ventilators (2 pcs)	15	Grinder in action and a ventilator	87.5	93.2	106.5

Note: Measuring equipment used and data measured: Integrated phonometer 2240 type. Accuracy level: 1, produced by: Brüel and Kjaer.

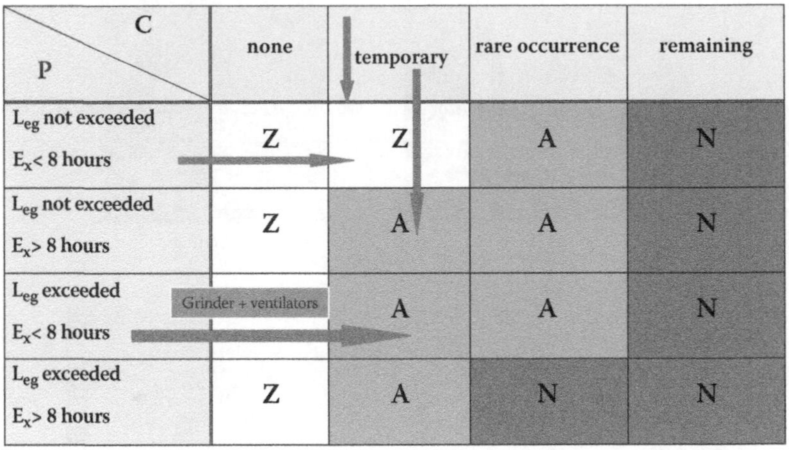

FIGURE 5.26 Application of acoustic risk matrix at laboratory.

A must be measured at a distance of 1 meter from the machinery surface and at the height of 1.6 meters from the floor or the platform; the location and maximum noise level must be stated (according to this measuring process). Noise reduction at the source is the most effective way of acoustic risk prevention and the possibility of performing the reduction should be included in the design of new machinery. When positioning machinery within the plant, it is essential for the location to be in compliance with Directive No.42/2006/EC to create an appropriate acoustic environment.

5.5.4 TOOLS FOR ACOUSTIC RISK MINIMIZATION

5.5.4.1 Research and Development

Scientific research on noise affecting the environment, methods for noise-level reductions, low-noise techniques, and development of special low-noise products are effective means of improvement and lead to reductions in acoustic risk. Funding pilot projects is useful for the purpose of presenting the advantages of technical and planning measures to decrease the effects of noise on people.

5.5.4.2 Legislative and Normative Documentation

Emission standards: Emission standards are defined by governments and determine emission limit values applied to individual sources and included in the type of approval procedures to identify whether new products are in compliance with the noise limits at the time of production.

Immission standards: Immission standards are based on the qualitative criteria or guide values for noise exposure that are to be applied at a specific location and are generally included in the planning processes.

5.5.4.3 Economic Tools

These are the types of economic measures that are or could be used within the policy of fighting noise; they include tax payments and noise emission payments, economic stimuli as the motivation to reduce noise and develop low-noise products, and compensation payments to those exposed to noise.

5.5.4.4 Operational Procedures

Among the most frequent measures are operation-friendly procedures to restrict the use of noisy products, machines, equipment, and vehicles in sensitive locations and at sensitive hours.

5.5.4.5 Room Acoustics Simulation Using 3-D Programs

Nowadays it is possible to design acoustic environment and room acoustic simulations using various programs, as such CATT Acoustic, Izofonik, Androl–noise 1.0.

5.6 RISK MANAGEMENT IN THE PROCESS OF MACHINE AND MACHINERY DEVELOPMENT AND DESIGN

Safety is the ability of a subject (machine, machinery, or products in general) not to cause harm to people or to seriously damage (destroy) materials or the environment. The following states must be taken into account in risk (safety) management:

- The state of a machine or machinery being in operation and performing activities in compliance with technical instructions.
- The state of a machine or machinery being out of operation and not performing activities in compliance with technical instructions.

The first state is related to prevention of malfunctions and, thus, against potential injuries (accident prevention), as governed by national and international regulations. The latter state presumes the elaboration of in-house regulations to protect the system against the potential consequences of the malfunction.

Regarding the above, it is essential to distinguish between the safety of machines, machinery, and technical systems, risk control, and their reliability. Within the risk control system, the degree is analyzed at which a machine or machinery is in a condition that is not hazardous to the safety of people, operators, and third parties, that is, in a safe condition even in case of failure (fail-safe behaviour). The machine and machinery reliability analyses are aimed at defining their failure-free conditions, that is, reaching the lowest possible number of shutdowns. Moreover, within machine and machinery safety, technical safety, the effects of external factors, must be taken into consideration, such as human error, catastrophes, and sabotage (civic security factors). The measures designed to minimize risks of machines and machinery normally contribute to their increased reliability.

5.6.1 Current Trends in the Construction Design Process

A current trend in the area of product construction design is toward a purposeful systematic approach. The properties of the quality and safety of a final product are the crucial parameters of product development.

The construction design process makes use of a multistage concept of the technical system design worked upon by several designers of different expertise. The concept includes the contribution of safety engineers whose task is to achieve predefined parameters of the projected system safety using all the knowledge available and in compliance with valid legislation.

The focus of the systematic approach is not only on the preproduction and production stages, that is, the construction design and production technology design, but also on the issues of ensuing product recycling as a part of its technological lifespan and environmental safety.

One of the crucial product usage parameters is the consumer's satisfaction, which requires a manufacturer to create a high-quality, safe, and reliable product. Based on this, designing for safety is a systematic and methodical activity and utilizes a variety of standardized procedures in the process of creating a final product with partial objectives and the final objective being taken into account.

Designing safety includes not only the appropriate materials selection and correct sizing of a component to comply with a respective technical standard (EN, STN) but also the integration of the product safety requirements, conformity assessment, and subsequent product certification as essential parts of launching the product (Directive No. 2/2006/EC). The algorithm of applying these measures into the design process is shown in Figure 5.27.

The more economically developed the society is, the more attention is paid to safety issues. The Slovak Republic as one of the EU member states treats the issues according to Act No. 264/1999 Coll. on Technical Requirements for Products and on Conformity Assessment and Government's Decree No. 436/2008. The given enactments are further amended by the Slovak government's decree on technical requirements for machinery.

Machinery must be designed so its operation, settings, and maintenance are not harmful to people and property, provided that the machinery is used under the expected conditions. The aim of the measures applied must be the elimination or reduction of risk of any sort of injury occurrence during the expected lifespan of the machinery, or of property destruction during the lifespan, even under operating conditions that are not in compliance with the technical requirements for operation.

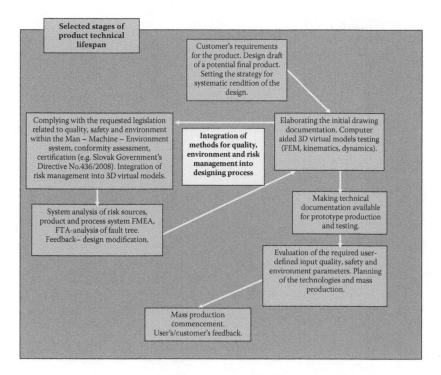

FIGURE 5.27   Integration of risk management methods into machine design process.

Current trends in construction design lean toward the more frequent utilization of computer-aided (CA) techniques. The area has recently been labelled as *virtual prototype* (VP) technology. The requirements for final construction design are aimed at utilizing available tools of technical documentation, which has resulted in increased productivity and product quality, the decrease in product development costs and reliability growth costs, and the reduction of development time (see Figure 5.28). In accordance with the above-mentioned enactments, it is crucial for safety, as a part of the final quality of a product, to be directly integrated into construction design.

Moreover, the current trends of construction design utilize more intensively CA techniques. CA techniques have become an essential part of analyses in the field of (technical) safety, quality, and reliability predictions. In that connection, the following are some CA procedures that have become part of modern engineering activities:

- Computer-aided design (CAD)
- Computer-aided engineering (CAE)
- Computer-aided planning (CAP)
- Computer-aided manufacturing (CAM)
- Computer-aided quality assurance (CAQ)

For the purpose of an integrated approach in the production process, the CA systems are organized in so-called computer-integrated manufacturing (CIM) or computer-aided industry (CAI).

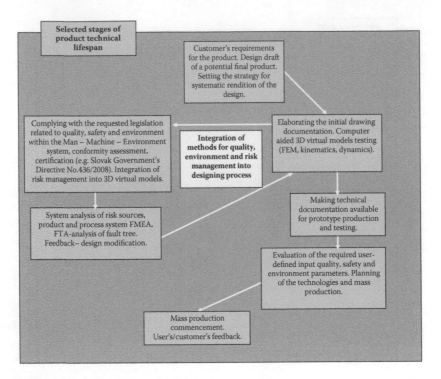

FIGURE 5.28 Comparison of design system development trends.

Most changes in a construction design occur at the initial development stages with the possibility of the construction designs to be realized in a less complicated and cheaper way. The final (ultimate) result includes:

- Production of fewer prototypes
- Lower production costs
- Shorter product development period
- Construction designs of high quality following the risk management principles in compliance with the EU States legislation (Figure 5.29)

The systematic approach to the implementation of CA techniques within concurrent engineering (Figure 5.30) is currently applied at a majority of designer's offices and also by a vast majority of middle-sized manufacturers, not to mention major corporations, such as automobile manufacturers, where implementing CA techniques has dramatically reduced the innovation cycle and increased the final product quality (Ford, General Motors, Nissan, Boeing, Volkswagen).

5.6.2 CA TECHNIQUES: RELATED RISK

The integration of computing programs and CA techniques into the construction process has recently been among the greatest changes in designer's work. The extent of

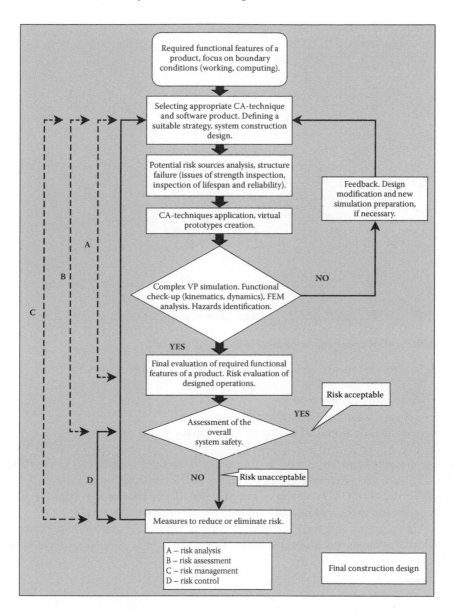

FIGURE 5.29   Risk management scheme at the machine design stage.

CA technique application is dependent on the type of products and industries. The requirements for reducing the time periods for development and realization of new products, as well as a growing number of variant products or their modular conceptions, accelerate the implementation of CA techniques at firms of all industries. Nowadays, it is hard to imagine the common machine construction designs or projects involving complex machinery being realized without CAD methods being applied.

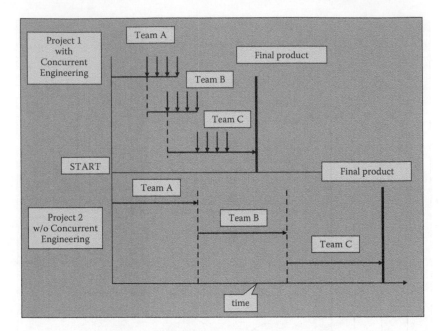

FIGURE 5.30 Concurrent engineering in the design process.

Apart from technical, organizational, and economic prerequisites for

- faster work completion,
- systematization of repeatable tasks, and
- reducing the workload as a result of schematic works,

CA techniques bring about new, and for many companies indefinable, risks. These days, if there are systematic errors occurring in the construction process, then the implementation of CAD techniques and, thus, the occurrence of new potential hazards contribute to the multiplication of the number of possible errors.

CAD programs and their application comply with the expected requirements, provided that the technical specifications of computers make possible a simple, complete, and correct

- integration of all relevant and valid legal regulations related to the technologies,
- implementation of the latest scientific developments, and
- application of corporate experience and knowledge of all creative staff.

The systematic addition to the standard software of field-oriented program libraries, additional program modules, and experts' experience transformed to software products is of significant importance.

The program libraries and additional modules are suitable for use in designing and projection processes only after their features have been inspected by experts and are complete with specific subprograms applicable to individual cases of their own production program. CAD programs today include normative calculations according to the respective standards with the emphasis put on the accessibility by a user, for example, a designer or calculator. The process of defining a product's reliability often requires fulfillment of valid standards. The problematic issue is usually caused by the fact that the standards have been created based upon simplified conditions and the required safety (risk reduction requirements) is ensured via a variety of corrective factors determined upon testing. On many occasions it could be objected that the standards concept does not match the potential of modern computing methods, for example, the finite element method (FEM). The standards, however, include engineering experience and instructions for the safe design of standard cases. It must be admitted that strict standard implementation within the creative design process causes a reduction in creativity, and in many instances, results in the final design lacking the characteristics of an optimal solution!

According to their content and use, additional control mechanisms could be integrated into CA techniques. At every stage of a design process, the mechanisms evaluate the applications that are required by the integrated approach to new designs of machinery, that is, the requirements for safety, quality, and environment.

The problematic area is the area of systematic updates of new scientific findings in respective fields, and thus the extension of program packages for CAD techniques. The updating process cannot be carried out by a software company; it must be done by a group of experts active in the given field of expertise. The specific technical data from new generally binding regulations related to technology, standards, recommendations, public notices, and decrees, not found in charts and sheets, must, for the sake of practical usage, be:

- Modified in the form of detailed technical clarifications
- Modified to be directly integrated into the draft versions of CAD programs
- Modified in the form of database entries for the purpose of modular variant solutions

It is essential for the data to be updated in accordance with the latest scientific developments. The update may be performed only by experts from particular industries.

The risks of working with CAD systems is of a vast nature and is dependent on the CAD system user's work as well as on the quality of the CAD program itself.

CAD system risks can be divided into several categories, as shown in Figure 5.31.

5.6.2.1 Controllable and Partially Controllable Risks

These risks can be partially or completely reduced by applying the systematic approach to CAD systems implementation.

The process of CAD systems implementation begins with required CAD system selection. It is not possible to say with certainty which CAD system is the most appropriate. The selection process should include the final implementation of the CAD system, the limits (often of an economic nature) of the company/designer's office related to the degree of complexity of the CAD system required, since each

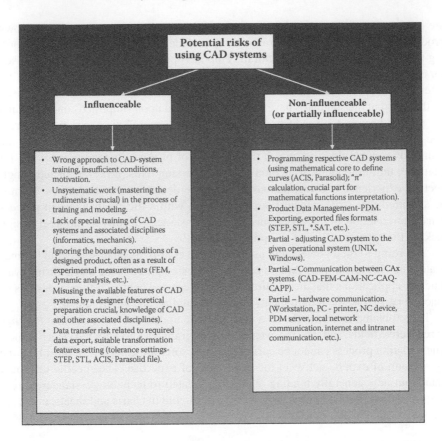

FIGURE 5.31   Overview of risk sources when using CAD systems.

degree of software performance capacity calls for appropriate hardware capacity, which eventually has an effect on the efficiency of the CAD systems implementation.

The educational process is yet another factor that affects the final efficiency of CAD program implementation, and in addition, the ensuing levels of risk influence. The system of *self-education* is to be accepted only at the level of a small-sized design company (few designers). Provided that a more complex CAD systems implementation is planned for (tens of designers), systematic education by means of courses as part of further education must be considered. The cooperation (bonds) between companies (CAD systems users) and universities and other educational institutions is advisable; university-educated designers would be given the rudiments of designing and working on e-construction of machines, complex machinery, and technologies.

As far as CAD systems application of risk reduction is concerned, it is important for a user (developer or designer) to be familiar with other scientific disciplines, such as mathematics, descriptive geometry, elementary technical mechanics, mechanical elements, and other related disciplines, including electronics, IT techniques, and technology with regard to mechatronic principles of modern machines and machinery. It is vital to understand that CAD systems are merely tools for creating a construction design and are not 'cure-alls' for problem solving.

A frequent source of risks when working with CAD systems is inappropriate consideration of boundary conditions (input parameters) of the working environment of a future technological system or product during the creation of its virtual prototype. Regardless of the perfection of the model, when loaded with inappropriate boundary conditions, the model becomes merely a *precise calculation of imprecise numbers*. This results in a model not matching reality, and the advantage of using a CAD system is lost. These kinds of risks may be prevented by means of a suitable theoretical background and correct definition of the input parameters (often as a result of experimental measuring).

Further, partially controllable risks include data transfer between CAD systems. Although a designer cannot affect the type of exported or imported file, by means of parameter settings that are required for a particular file type such as defining tolerances, sizes, and shapes of items used, with, for example, the FEM, the designer may prevent potential risks of imprecise data transfer (3-D model enhancement) or data loss during transfer.

Data formats such as DXF, IGEC, SAT, SET, VDA, and VRML, supported by major CAD systems development companies, allow for the transfer of geometrical data only. The issue of non-geometrical data transfer is often dealt with by using additional modules, extensions, and tools created in various programming languages and implemented in CAD systems (e.g. AutoLisp in AutoCad system). Alternatively, external procedures are performed, normally in Delphi, C++, and Visual Basic, and those record mostly non-geometrical data into internal, purpose-created files. This solution is not standard and calls for advanced programming skills and a detailed understanding of the CAD system internal structure and internal database.

Similar problems occur between other CA systems that aid various activities within a company (engineering analyses, production planning and control, OHS management, quality management, assembly, etc.).

Another potential risk source related to CAD systems implementation in practical life is data management (product data management [PDM]). PDM as a data management system is essential in the process of CAD systems implementation in the automobile and aviation industries, architecture, and electrical industries. Companies dealing with CAD systems development pay close attention to the data management issues since, apart from high-quality data storage, a future user often opts for a CAD system for better data management.

5.6.2.2 Non-Controllable Risks

From the point of view of a user or designer, non-controllable risks connected with CAD systems application into a construction process are caused by the CAD system itself, that is, the CAD system environment and its programming. This area cannot be modified, since it is set by default by the CAD system manufacturer.

Modern CAD systems make use of various mathematical cores to calculate parametric functions of curves. ACIS and Parasolid mathematical cores are among the most frequently used. These CAD systems bases have been through a long-term development gradually enhancing their features (issues with drawing, e.g. curved areas and a variety of mathematically common areas, calculating intersections, 3-D bent curves, etc.).

Regarding the final inaccuracy of geometry calculation, as well as the FEM analysis outcomes, potential risk sources are represented by, for example, the definition of the 'π' calculation within mathematical operations. On some occasions, the discrepancies of the calculation results are not negligible.

The principle of calculating several continuous mathematical functions in computing programs is based on data discretization followed by interpolation of function values. The continuous function interval division into the finite number of points and the final function itself are further interpreted by the interpolar integration of the interval points function values. The type of interpolation may often have an effect on the optimization analysis outcome (e.g. FEM) or on CAD system geometry interpretation. In transformation, the different mathematical interpretations of the basic curves definition may cause geometry deformations or data loss, for example, defining a circle curve.

All these potential risk sources can possibly be reduced only partially by means of CAD technique setting options as defined by the software development companies. Feedback from CAD technique users and software development centres is of significant importance.

Eventually, the resulting degree of risk by using CA techniques is the difference between product features as defined by a designer and real, experimentally measured product features. It is in every designer's interest to reduce this risk degree to the lowest levels possible.

The essential part of reducing the risk of CA technique implementation into the construction design process is the feedback from experimental testing, that is, the mutual connection between teams of designers and final product users' experiences. The reduction of delay time and providing user's information on product features are some elementary means of CA technique risk minimization in a construction process.

5.6.3 Examples of Potential Risk Sources

Applying CA techniques to a machine construction design using a *virtual prototype* may create potential sources of risk in the construction design, provided that a designer lacks knowledge of the technique's proper application.

The method is based on the principle of division (netting) of a 3-D model into a finite number of elementary parts with predefined physical qualities (the analysis boundary conditions). In case of more complex geometry, even a sophisticated automatic netting of current FEM programs will not be able to correctly net out a 3-D model, which results in the elements being of significantly deformed shapes (see Figure 5.32). The FEM analysis outcomes will be inaccurate and not match the actual situation. Abolishing this kind of a potential risk source by a designer requires knowledge of the FEM analysis application.

It is obvious that a CAD system's user/designer is exposed to considerably vast sources of potential risks. The designer is often required to pay close attention to the risks, which causes the designer to overlook the core principle of his work, the construction design of a product. In order to reduce these potential risk sources, it is essential for risk control methods to be implemented directly in the CAD systems, as it is also necessary to cover the issues of final product safety and quality in the

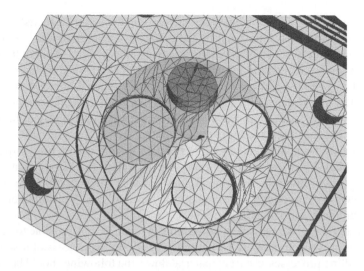

FIGURE 5.32 Deformed elements of a cylinder head; potential risk source.

form of a parallel program package for this to be used when working with the CAD systems and at every stage of product development.

5.7 RISK MANAGEMENT WITHIN MATERIAL FLOWS

Modern procedures of material flow projection within complex logistic procedures require the integration of risk management systems with the aim of creating safe material flows.

Directive No. 89/391/EC, on the introduction of measures to encourage improvements in the safety and health of workers, defines the employer's commitment to perform risk analyses at a workplace (Section II, Article 6, Paragraph 2, 3) and notify workers of same (Section II, Article 10, Paragraph 1). The enactment defines principles of occupational health and safety with the main principles being the commitment of designers, project engineers, and users of machines and machinery to assess risks.

Modern trends in the field of material flows projection aim at a purposeful systematic approach. The quality, safety, and reliability issues are crucial input parameters at the projection stage of a material flow system.

Customer's satisfaction is one of the most important parameters of using material flows; satisfaction requires a manufacturer to create a high-quality, safe, and reliable material flow (MF) system.

Machines and machinery systems within a material flow system must be designed so that their operation, settings, and maintenance, when performed in compliance with conditions defined by the technical documentation, are not hazardous to people and property. The aim of the measures implemented must be the elimination or reduction of risk of any negative event occurrence during the machinery technological lifespan, including assembly and disassembly, even under operating conditions that are not in compliance with technical requirements for the operation.

The measures designed to increase safety, and thus minimize risks within a material flow, are often influenced by the reliability of its individual components (transport machinery). Quantitative assessment of safety is carried out in the form of risk estimation or calculation; the acceptable risk level is defined by the value of its acceptability for individual machines and machinery types. Using risk for the purpose of safety assessment of the material flow requires interdisciplinary procedures. It is crucial to correctly define a relationship between the probability of failure or a negative event occurrence and the consequence that the failure or negative event may cause. It is also essential to take into consideration various causes of negative events within the Man–Machine–Environment system as well as their consequences (location, time, person involved, duration of consequences, etc.). Appropriate statistical methods should be used for the purpose of risk (especially the probability of negative event occurrence) estimation or assessment.

When applying risk assessment procedures in practice, it is possible to define the goal that the given activities are performed to fulfill. In the process of assessing the risk of handling processes as part of material flows, the following should be the focus of analyses:

- Profession (job)
- Workplace
- Complex material flows

5.7.1 Risk Assessment for a Certain Profession

Risk assessment related to a profession, that is, to activities performed by operational staff for a certain technical equipment within a complex material flow, for example, crane operator, or assessment related to performing the precisely defined activities within a material flow, for example, a lifting machine's maintenance person, is based on the assumption that the job is performed constantly and repeatedly and is connected to a single defined workplace (a cabin of a crane) or is performed at a variety of workplaces with an approximately same job description (a maintenance person performs similar activities on various types of cranes).

Risk analysis in accordance with EU Directive No. 89/391/EC includes a requirement for an employer to respect employees' requests. It is based on the assumption that the risks typical for a machine or machinery system utilized to perform activities in a material flow, or the tools used for the activities, are defined by their designers via the manufacturers. The users are thus informed by suppliers and employers about their existence and are the people who must be able to recognize the existence of residual risks. For the purpose of hazards assessment, as a part of a risk management process, the conditions occurring at the Man–Machine–Environment interfaces must be taken into account. An example of how to perform the analysis is the usage of catalogue sheets (see Table 5.4). This procedure fully complies with Directive No. 89/391/EC. The proposals for risk minimization processes are included in the ensuing activities as stated on separate sheets. This procedure has been selected because the analysis is performed by a team of experts whose goal is

TABLE 5.4
Example of a Catalogue Sheet

Plant: Page:
Operations: Date:
Profession: Done by:

Danger	Hazards	Hazard Type (e.g. by EN STN ISO 12100.1 and 2)	Related Regulations and Standards for Given Hazard Type

to identify and assess risks. Measures for risk minimization are included in particular standards and recommendations and their execution is normally carried out by another team of experts.

The identification of potential hazards associated with a profession or an activity, is important in any workplace (machine, equipment) where the duty is performed. The other characteristic feature is the validity of regulations that give an exact definition of the activities and their execution within a respective profession or a set of performed activities (e.g. work and safety regulations). Regarding this, it is possible to assess whether the activity description within a profession is in compliance with the regulations that govern these activities, for example, a crane operator on a travelling crane at a particular specific operation (Figure 5.33).

In the case of a single job being performed on various machinery types, for example by a crane maintenance worker, it is advisable to conduct the assessment of hazards related to the maintenance worker's activity at the workplace. The assessment is aimed at all the working operations that must be performed during the maintenance process regardless of the type of crane. Certain activities are identical for a group of lifting machines, as some sets of activities are typical for particular crane types. It is necessary to realize that the assessment process of activities actually performed in compliance with the given regulations is time-consuming. Typical examples this approach may include material flows within the engineering and construction technologies and gas industry, that is, in the areas where several people perform the same activity as a part of their job (Figure 5.34).

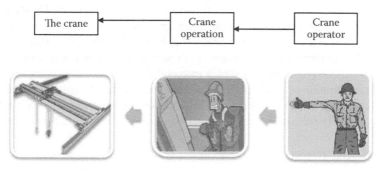

FIGURE 5.33 Risk assessment for a job.

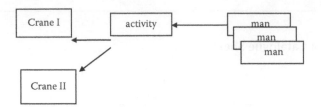

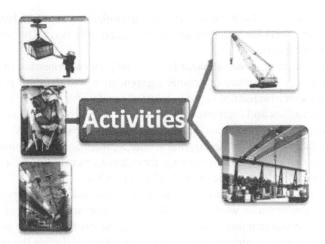

FIGURE 5.34 Example of a single job done by several employees.

5.7.2 WORKPLACE RISK ANALYSIS

This procedure is to be applied if there are various activities performed at a workplace to ensure correct operations within material flows and, thus, operations involving machines or complex machinery. A variety of activities are assumed to be performed at a single workplace, such as crane operations as the main working activity with other activities (e.g. crane maintenance) supporting the main activity. Unlike the two above-mentioned approaches in which a worker is an active element, this approach considers a worker as a part of the system, that is, the character of working activities at a workplace determines the jobs that are required for the particular activities to be performed. At the projection stage of a material flow, the necessary activities for a failure-free operation of machines and equipment are defined, which jobs must contribute to this, and what the effect is of machines on the surrounding area in which a worker is a part of the Man–Machine–Environment system (Figure 5.35).

5.7.3 PARTIAL CONCLUSION

When selecting a risk analysis method, several procedures can be used. The procedure and the risk assessment method selection are dependent on the goal to be

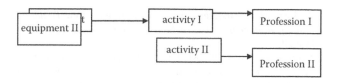

FIGURE 5.35   Risk analysis for a workplace.

achieved. Each of the procedures described could lead to an effective result provided that it is carried out by a team of experts in the field of occupational health and safety, of experts coming from the operation being analyzed, and of employee representatives. It must be realized that the analysis outcomes must create conditions for risk minimization. Provided that the risk analysis is applied at the stage of the material flow system projection, the outcomes must enable designers to implement risk minimization measures at the projection stage. If this is impossible or uneconomical, the residual risks must be stated in the technical instructions or user's manuals for individual types of transporting machines and handling equipment within the defined material flow system. At the same time, the measures are designed to be implemented when the machines are put into operation within the material flow systems.

BIBLIOGRAPHY

'Einige Überlegungen zur Risikoanalyse während des Kranbetriebes', in *Der Kran und sein Umfeld in Industrie und Logistik*: 19. Internationale Kranfachtagung Magdeburg, ILM, 2011, pp. 119–125, ISBN 13:978-3-930385-74-4.

'Manažérstvo rizika ako predpoklad pre voľbu stratégie údržby', in *National Maintenance Forum*. Žilina: NMC, 2000, pp. 102–105, ISBN 8085655152.

Miskolcer Gespräche. *Die neueste Ergebnisse auf den Gebiet Fördertechnik und Logistik*, Miskolc, September 2003, pp. 1–5, Tagungsband ISBN 963 661 595 0.

Pačaiová, H., and Sinay, J. 'Integrovaná bezpečnosť strojov: Jej prínos v údržbe' (Integrated safety of equipment: Its contribution in the maintenance), in *ÚDRŽBA 2004*. Praha: Česká zemědělská univerzita v Praze, 2004, pp. 173–177, ISBN 8021312114.

Pačaiová, H., Sinay, J., and Glatz, J. *Bezpečnosť a riziká technických systémov*, edited by SjF TUKE Košice, Vienala Košice 2009, ISBN 978-80-553-0180-8-60-30-10.

Sinay, J., and Balažiková, M. 'Acoustic risk management', in *Human Factors and Ergonomics in Manufacturing* 20, no. 4, pp. 329–339, John Wiley & Sons, Inc. Malden, USA, 2011, ISSN 1520-6564.

Sinay, J., and Cahajlová, M. 'Metodika riadenia rizika hluku', *New Trends in Occupational Health and Safety Management and Technical Systems Safety*. Košice: TU-SjF, 2002, p. 5, ISSN 1335-2393.

Sinay, J., and Cahajlová, M. 'Riziká v dôsledku akustického tlaku v podmienkach strojárskej praxe' (Risk due to acoustic pressure in the conditions of mechanical engineering practice), in *DOKSEM 2002*. Žilina: Žilinská univerzita, 2002, pp. 13–18.

Sinay, J., Ferenčíková, A., and Vargová, S. 'Checklist pre bezpečnostný riadiaci systém petrochemického podniku zaradeného do kategórie B podľa zákona 261/2002 Z.z. o prevencii závažných priemyselných havárií', in *Topical Issues of Work Safety: 22nd International Conference: Štrbské Pleso-Vysoké Tatry*, 18–20 November 2009. Košice: TU, 2009, pp. 1–5, ISBN 978-80-553-0220-1.

Sinay, J., and Laboš, J. 'CAD-Techniky a ich riziká pri konštruovaní strojov I', *Strojárstvo*, March 2000, pp. 13–15, ISSN 1335-2938.

Sinay, J., and Laboš, J. 'Možnosti integrácie požiadaviek kvality a bezpečnosti do procesu konštruovania', in *Collection of Abstracts at 26th International Transport and Handling Departments Seminar*, STU Bratislava, 2000, pp. 94–101, ISBN 8022713759.

Sinay, J., and Laboš, J. 'Možnosti integrácie požiadaviek bezpečnosti do etapy vývoja a konštrukcie strojov', in *Topical Issues of Work Safety*. Bratislava: VVÚBP, 2000, pp. 49–58.

Sinay, J., and Laboš, J. 'Integrovanie metód riadenia rizika do procesov konštruovania stroja' (Integrating of risk control methods into machine construction process), in *Equipment Quality and Reliability*. Nitra: SPU, 2000, pp. 93–96, ISBN 8071377201.

Sinay, J., and Laboš, J. 'Analýza rizík aplikácie CAD systémov ako súčasť bezpečného konštruovania I', *Strojárstvo*, August 2001, pp. 32–33, ISSN 1335-2938.

Sinay, J., and Laboš, J. 'Analýza rizík aplikácie CAD systémov ako súčasť bezpečného konštruovania II', *Strojárstvo*, September 2001, pp. 80–81, ISSN 1335-2938.

Sinay, J., and Laboš, J. 'Analýza rizík aplikácie CAD systémov ako súčasť bezpečného konštruovania III', *Strojárstvo*, October 2001, pp. 52–53, ISSN 1335-2938.

Sinay, J., and Laboš, J. 'The risk analysis of the application of the CAD systems as a part of safety construction', *International Conference Micro CAD 2001*, University in Miskolc/Hungary, pp. 89–94, ISBN 963 661 457 1 (963 661 468 7).

Sinay, J., and Laboš, J. 'Application of knowledge-based systems in risk analysis of CAD systems', 8th International Conference on Human Aspects of Advanced Manufacturing, Roma, National Research Council of Italy, 2003, pp. 571–575, ISBN 88-85059-14-7.

Sinay, J., Majer, I., and Hoeborn, G. 'Risk in mechatronics systems', XVIII. World Congress on Safety and Health at Work, June 29–July 2, 2008, Seoul Korea, Sektion 26. Sicherheit von High-Tech Kontrollsystemen übernehmen die Führung bei der Sicherheit am Arbeitsplatz.

Sinay, J., Malindžák, D., Pačaiová, H., and Malindžák, D. Logistics principles application in MIA (Major industrial accident) prevention. Miskolcer Geschpräche, 2003, Miskok, Hungaria, 2003, pp. 1–5, ISBN 963-66159-5-0.

Sinay, J., Oravec, M., and Majer, I. 'Beurteilung des Risikos im Mensch–Maschine–Umwelt', *System International Conference Globalna Varnost Collection of Abstracts*, ZVD Ljubljana, Bled, Slovenia, June 2000, pp. 19–27, ISBN 961-90350-7-0.

Sinay, J., Oravec, M., and Pačaiová, H. 'Údržba: Prostriedok pre ovládanie a znižovanie rizika', Conference: Operational Reliability of Production Equipment in Chemical and Food Industries, Slovnaft a.s. Vlčie hrdlo Bratislava, October 1997, p. 10.

Sinay, J., and Pačaiová, H. 'Logistika a riziká spojené s jej realizáciou v praxi', Conference: Transport, Material Handling, Logistic systems, Dum techniky Ostrava s.r.o., 1997, pp. 47–52.

Sinay, J., and Pačaiová, H. 'Logistika a rizikové faktory', Conference: Transport and Handling in Logistics, Výstavisko TMM, Trenčín, 1998, pp. 7–13.

Sinay, J., and Pačaiová, H. 'Sicherheitskriterien in der Etappe des Projektes von Materialflussysteme', *International Conference Miskolcer Gespräche 2001*, University in Miskolc, pp. 1–7, ISBN 963 661 493 8.

Sinay, J., and Pačaiová, H. 'Analyse und Bestimmung der Risiken im Hubwerk eines Brückenkranes 10', *Internationale Fachtagung 2002 Kranautomatisierung, Kompenente, Sicherheit im Einsatz*, Magdeburg, IFSL Otto-von Guericke Unuiversität Magdeburg, Reihe III: Tagungsberichte Nr. 16, June 2002, pp. 31–43, ISBN 3-9303385-37-6.

Sinay, J., and Pačaiová, H. 'Úlohy a ciele manažmentu rizika a manažmentu údržby', in *National Maintenance Forum 2002*. Žilina: ŽU, 2002, pp. 32–37.

Sinay, J., and Pačaiová, H. 'Risikoorientierte Instandhaltungsstrategie', TÜ Bd. 44 (2003) Nr. 9, VDI – Verlag Düsseldorf, September 2003, pp. 41–43, ISSN 1434-9728.

Sinay, J., Pačaiová, H., and Kopas, M. 'Maintenance and risks during maintenance operation 1', *International Conference on Occupational Risk Prevention, ORP 2000*, Tenerife (Canary Isles), Spain, February 2000, ISBN 84-699-1242-9, CD.

Sinay, J., and Šviderová, K. 'Neue Risiken auf Grund aktueller demografischer Entwicklungen', in *Forum Prävention 2011*, 9—12 Mai 2011, Kongresszentrum Hofburg, Wien—AUVA, 2011, pp. 1–6.

Sinay, J., Šviderova, K., and Tompoš, A. 'Vplyv zraku na bezpečnosť a ochranu zdravia pri práci', in *Conference: Occupational Health and Safety 2010*, VŠB-TU Ostrava, MPaSV ČR, May 2010, pp .251–261, ISBN 978-80-248-2207-5.

6 Risk Management and Its Application in Safety and Security Systems

Currently, technical and technological equipment, such as complex machinery systems, as well as individual machines, devices, and tools, are required to meet high standards set by the market. Today, the features of these subjects are viewed in a broader sense and include the requirements stated in Table 6.1.

The following basic rule is applied:

> In order to prevent a fault or accident (injury), it is necessary to promptly identify possible risks and activate respective countermeasures.

In relation to the implementation of effective measures within the process of risk minimization, in the field of occupational health and safety (OHS), in technical equipment safety, as well as in the area of civil security, it is necessary to become familiar with the actual conditions of the technical equipment that is a part of the Man–Machine–Environment system. During the operations of technical equipment, it is necessary to determine its actual technical conditions through the collection of data that characterize these operational conditions.

6.1 MAN AS THE OBJECT OF SAFETY ANALYSES WITHIN SECURITY AND SAFETY

Within the Man–Machine–Environment system (see Figure 6.1), it is the human factor that plays a crucial role in both safety and security management.

The safety system deals with consequences that are the result of unplanned activities, and that on many occasions, occur as a final product of man's failure, that is, the human factor.

Within the security system, the term *dangerous human* may be used. The goal is to create a causal relationship of negative event occurrence so that the period between losses is as long as possible.

Regarding the overlap of activities in safety and security management, it is important to focus on the integral evaluation of the human factor role in relation to achieving the ultimate goal—the protection of all the elements in the Man–Machine–Environment system.

The activities of a fire brigade can serve as an example. Firemen contribute to the elimination of the effects of fire, which falls in the area of security. When

TABLE 6.1
Features of Technical and Technological Equipment

Functionality	Fulfilling the defined tasks at requested quality
Meeting the delivery deadline	Ability to deliver a product promptly and within a required period of time
Efficiency	Efficient production and subsequent operation of the equipment
Long lifespan	Sufficiently long and safe operation
Environment protection	No illegal environment pollution in the process of production, operation, and re-evaluation
Occupational health and safety (Safety)	Creating safe work environment and/or technology
Safety of technical equipment (Safety)	Safe machine designs, machinery systems, and complex technical and technological equipment
Civil security (Security)	Safe conditions for people's everyday life
Transport	Transporting the equipment to the final destination safely and at limited cost
Competence	Ability of the operational staff to utilize the operational tools
Maintenance	Keeping safe during maintenance and using safe spare parts and materials

FIGURE 6.1   Man–Machine–Environment system.

performing their activities, however, the firemen must be protected so they can carry out their tasks safely (safety), in order to prevent health hazards. Neglecting the firemen's occupational health and safety would cause an inability to carry out activities as a part of security, and thus an inability to ensure the protection of other people. It is important to pay attention to decreased psychological pressure during activities within the security system, which stems from the character of the activities, for example, saving human lives. Under these conditions, the firemen's weakened ability to follow all the safety conditions necessary for their work duties may be weakened, for example, the perception of requirements for health and safety.

In order to increase the efficiency of performed activities and minimize risks, the following approaches could be used:

- Intensive and effective training at the stage of activity performers' preparation (within security) until automated routine is achieved
- Developing hardware and software tools to aid in the process of minimizing human errors when carrying out job activities, regardless of the human factor influence

Creating effective conditions for minimizing losses for both the safety and security systems is built on common ground. In both cases, it involves looking for opportunities, ways, and methods to interrupt the causal relationship of the negative event occurrence. These procedures must be included in scientific and research works of individual countries and, especially in the area of security, also of international research consortia. The importance of this field of research as the main focus of the European Union has been underlined by the fact that the security area has been included in the science and research priorities, for example, within 7th FP EU for 2007–2013.

Safety and reliability of a machine or equipment must be systematically monitored and inspected and there must be conditions created to minimize negative events in the form of interruption of the casual relationship of failure, accident, and injury occurrence, that is, minimizing the risks. A technical equipment operator must be familiar with the methods and tools for performing these activities, and there must be a sufficient operational structure in place for the machines and equipment to be operated in a safe way. One of the effective means of carrying out these activities is the collection of processes within maintenance technology and technical diagnostics as its part.

Based on the given facts, all the effective maintenance processes are applied to provide conditions for the safe and reliable operation of machinery systems, technologies, and individual machines during their lifespan, that is, to ensure a constant ability to apply measures to minimize technical and human risks.

Risk minimization requires application of processes that can identify the failure occurrence potential as early as at the hazard stage. Modern and effective maintenance methods are characterized by their interdisciplinary nature and also include, to a large extent, modern technical diagnostics methods.

Technical diagnostics is called *diagnostics*, that is, the activity that leads to the assessment of the object's actual conditions. Attention is paid to technical objects, such as machinery systems, complex technologies, and individual machines.

Experts in the field of technical diagnostics focus their attention mostly on activities in the area of prevention of failure, accidents, and injury occurrence, which is among the most effective activities from the point of view of human and technical safety, as well as from a financial point of view. The means of creating the conditions for minimizing risks (decreasing the frequency and effects of failures, accidents, and injuries) include the application of technical diagnostic methods in a variety of industrial activities, this being the most effective means of prevention.

Knowing the actual conditions of technical equipment enables its user to compare the actual operational characteristics with those defined by the producer. The emphasis here is on the ability to promptly apply all the measures to ensure that the equipment is operated the way the user expects. As a result of faulty technical subjects, the health of those operating the machine, or those within the area affected by the fault, may be in jeopardy; in other words, the principles of occupational health and safety management (safety), as well as those of civil security (security), are not followed.

During the operation of technological equipment, it is necessary, in order to evaluate its actual technical conditions, to collect data that characterize operational conditions, that is, to apply technical diagnostic methods in alignment with the latest methods of experimental measurement. Furthermore, it is important to verify the data measured as well as the data processing methodology to eliminate their irrelevance to the monitored subject.

6.2 NOTES ON THE DEFINITIONS OF SAFETY AND SECURITY

6.2.1 Safety (Occupational Health and Safety and Technical Systems Safety)

The meaning of the word *safety* stems from the French '*sauf,*' meaning 'no injury.' The emphasis is on assuring the state of safety, the state in which the risks of all the elements of the Man–Machine–Environment system are eliminated, with a focus on the human factor and technical subjects.

Safety is defined by a set, form, and interconnection of measures for minimizing physical, social, financial, mechanical, chemical, psychological, and other types of risks (risk = hazard potential), thus creating the system of activities for risk prevention and protection of human and material values in society. The occurrence of damage and loss is a result of unintentional negative interaction within this system.

Example 1: Aircraft safety, which includes the system of measures to minimize or eliminate risks generated by, for example, a wrong decision made by a pilot when operating a plane, the malfunction of the safety device onboard, or the errors of the ground crew. One of the possible effective measures to minimize this kind of fault is the application of redundant systems within airplane control systems.

Example 2: As ordered by law, the energy generation process at power plants includes the obligation to comply with strict regulations for risk elimination related to the activities with an output, electricity, with the in-house safety principle taken into account as early as the projection stage, as well as at the design and commencement stages.

Example 3: Information and communication techniques and technologies must be designed so they control, monitor, and maintain technologies to minimize risks—the tool for implementation of effective measures.

6.2.2 SECURITY (PROTECTION OF CITIZENS: CIVIL SECURITY)

Security is defined as a system of measures, forms, and methods for protection against destructive effects on human and material values. This is mostly initiated by man. Man affects an object, material, and environment with the aim to cause intentional loss through these subjects! Civil security also includes minimizing the consequences of risks that occur as a result of major industrial accidents, or negative events that may seriously threaten third parties, for example, the failure of energy distribution media (gas pipeline, oil pipeline, electricity distribution system, etc.).

Regarding natural disasters, there are processes that at times cannot be controlled (earthquakes, flooding, volcano eruptions, etc.), and eventually cause major human and material loss.

Example 1: Civil aviation security, which includes activities to protect air transportation from intentional destructive human action against equipment, the airport, and against people (e.g. a plane hijacking threat resulting in airport security inspections, luggage checks, etc.).

Example 2: In the process of planning a nuclear power plant, the possibility of the nuclear technology being sabotaged for the purpose of harming humans and the environment must be taken into account.

Example 3: Information and communication technologies could be used to cause predefined loss. Therefore, it is important to design such technologies so that they cannot be misused and so that they meet all the requirements of the technological environment.

6.3 COMMON ASPECTS OF SAFETY AND SECURITY WITH THE POSSIBILITY OF APPLICATION OF TECHNICAL DIAGNOSTIC METHODS

The safety and security areas overlap and influence one another in a variety of aspects and, as such, form a complex system of people's life, health, and property protection (see Figure 6.2).

The elements common to the safety and security areas include:

- The objects that they influence and that must be safe, and therefore protected, are mostly people (either workers—safety, or third parties outside a technical subject—security).
- The space in which the areas function (every company is physically a part of the state, the environment part of the Man–Machine–Environment system, and is governed by the state's legislation).

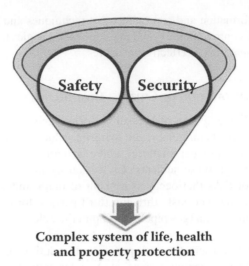

**Complex system of life, health
and property protection**

FIGURE 6.2   The model of safety and security integration.

- Prevention, the process of subject protection at a certain time and place, must be carried out so that risks are minimized at the danger and hazard stages.

Since the current notion of security in its broader sense is based on the principle of prevention, the areas of occupational health and safety, technical systems safety (safety), and civil security (security) often make use of identical means of diagnostics for the timely identification of the potential for the occurrence of a failure, accident, or injury, as the occurrence of an accident or injury of citizens outside the technical equipment or complex technology may be a result of a technical system failure.

In this case, these are the technical diagnostics that work to find possible failures that may further result in the technical equipment breakdown (e.g. a nuclear reactor and its parts, gas pumping station failure, or gas pipeline failure). The field of safety includes mostly methods and means of machinery and staff protection, which further minimize possible effects on people and property safety outside the technological unit.

Similar processes are also applicable in production, processing, packaging, and chemical substance usage. Substances that show at least one hazardous feature are included in the hazardous chemical substances list. The list includes, among others, radioactive substances that represent high risk to people but, on the other hand, are a crucial element in the process of electricity production.

6.3.1 Example: Nuclear Power Plant

Nuclear power plants must meet strict requirements especially in the field of safety and reliability. This is not a concern of a single country only; it is a multinational issue. Based on this, it may be stated that the means of technical condition diagnostics at nuclear power plants are an inseparable part of the safety system. As such, the means of technical diagnostics may be understood in two different ways.

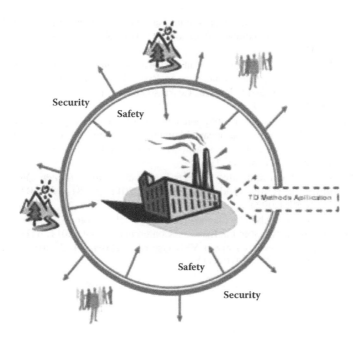

FIGURE 6.3   System of safety and security.

On one hand, there are the means of technical system diagnostics, for example, combined temperature sensors for determining the input and output temperature of the cooling media in the nuclear reactor. On the other hand, the means of the diagnostics at a nuclear power plant also include a detector for monitoring the amount of radioactive substances in the air (*dräger*) available for each staff member. The detector provides a timely warning, and as such, represents an individual diagnostic means that defines a potential hazard to people not involved. Both of the previously mentioned diagnostic means are utilized in practice simultaneously, which goes to show that the diagnostic system in the given environment uses redundant aspects, and thus increases the safety level (see Figure 6.3).

It follows from the given statement that there exists interconnectivity of the fields of safety and security with the methods of technical diagnostics that protect people not only in the OHS area or technical safety, that is, within the working environment, but also outside the working environment as a part of civil security, and the methods as such constitute a complex protection system that needs to be further improved and developed to minimize complex risks.

BIBLIOGRAPHY

Sinay, J. 'Niektoré poznámky k vzájomnému vzťahu Safety a Security', *Conference Occupational Health and Safety 2010*, VŠB-TU Ostrava, MPaSV CR, May 2010, pp. 244–250, ISBN 978-80-248-2207-5.

Sinay, J. 'Safety and Security in SR: Bezpečnosť pri práci a ochrana občana—synergie a presadenie do praxe', 23rd International Conference, *Topical Issues of Work Safety* and International Symposium, *Prevention in the EU 27- Focus SMEs*, ISSA,

International Social Security Association, Národný inšpektorát práce and Technická univerzita in Košice. September 29–October 1, 2010, Košice, 2010, pp. 89–94, ISBN 978-80-553-0481-6.

Sinay, J. 'Údržba a riziká—ich vzájomná interakcia v podmienkach Safety a Security', in Top Managers Summit on Maintenance in Risk Management, Conference Seminar, Liblice, April 13– 14, 2011, Prague: Česká společnost pro údržbu, 2011, pp. 27–36, ISBN 978-80-213-2172-4.

Sinay, J. 'Technická diagnostika a riziká—ich vzájomná interakcia podmienkach "safety" a "security",' in Central European Maintenance Forum 2011: 11th International Conference: Collection of Lectures: May 31, 2011–June 1, 2011, Vysoké Tatry, Štrbské Pleso. Žilina: ŽU, 2011, pp. 64–71.

Sinay, J. 'Security research and safety aspects in Slovakia', in *European Perspectives on Security Research*. Berlin, Heidelberg: Springer, 2011, pp. 81–89, ISBN 978-3-642-18218-1, ISSN 1861-9924.

Sinay, J., and Vargová, S.'Technická diagnostika a riziká, ich vzájomná súvislosť v podmienkach Safety a Security', 30th International Conference DIAGO 2011, ATD of the Czech Republic and Technická univerzita VŠB Ostrava, February 1–2, 2011, Rožňov pod Radhošťom/CR CD, ISSN 1210-311X.

7 Education as Part of the Training of Experts in Risk Management

Safety first. Vision Zero—vision of zero injuries. These are ideas with an unconditional priority in any part of social life nowadays.

Occupational health and safety management, or risk management, includes a summary of activities that follow one common goal—safety of the Man–Machine–Environment system. It was not many years ago that the experts' opinion of occupational health and safety being related first and foremost to the field of engineering was prevalent. Experiences and technological development only proved that it is not possible to replace the role of a man nowadays, and that the system Man–Machine will still be relevant. One of the goals of environmental management systems is the minimization of environmental risks and, finally, their elimination—all of the procedures included in the system of risk management.

We can comprehend risk management as an integration of several areas that have the goal of providing safety within the Man–Machine–Environment system. We can therefore deduce that risk management belongs in the science of safety. Not long ago, experts claimed that safety technology belonged to the engineering disciplines. Automation was expected to quickly assert itself in most of the field of engineering; however, development of technology clearly proved that there are plenty of fields where the role of a man cannot be substituted. We must also realize that the influence of human environment will be gradually taken into account by the Man–Machine–Environment system with more intensity. On the other hand, machines as well as people, separately or together, will be considered as a source of environmental risk.

Conventional forms of occupational health and safety originated from the assumption that they are extraordinary tasks for a smaller group of experts named by management. Responding to problems in the area of occupational health and safety, to injuries and accidents that happened, and to make sure that everyone followed the binding regulations were some of the main tasks. Paying attention to management systems, even in relation to occupational health and safety, influences the changes in these procedures. A swift change of the current labour market, which plays a critical role in gaining and transferring responsibilities, requires constant application of new forms and methods of occupational health and safety management.

It is necessary for managers and occupational health and safety management experts to not only adopt modern trends in the field of occupational health and safety, but to also lead by example in their work to achieve these goals. Their task is to influence integration and provide the execution of occupational health and safety rules, by providing information, communicating, lecturing, and by thorough inspection. It is expected that they will associate themselves with rules of occupational health and safety and that they will be open to the opinions of their co-workers. To make these activities happen, it is extremely important for them to study gradually in order to provide education for workers of any management competencies.

It seems that the current conventional disciplines and fields of study cannot cover the whole extensive topic of risk management, even though risks in the modern industrial society are a function of many parameters. An expert in the area of occupational health and safety should incorporate knowledge, experience, and maybe even skills of not only an engineer, but also an electrician, physicist, chemist, psychologist, sociologist, doctor, and even of other experts. However, it is only natural that one does not have the capacity to acquire, evaluate, and successfully put to use such an extensive database of information. Qualified experts in the area of occupational health and safety are therefore required to be distinctive in their ability to embrace, arrange, and prioritize new information to facilitate conditions for teamwork. This is conditioned by all of the modern legislative regulations on the level of the European Union; for instance, Directive 391/89/EC or 42/2006/EC exactly state the demand for integrating requirements of safety into features of products or production technologies.

For carrying out risk analysis, the directives offer the following:

- Directive 391/89/EC: Tools for risk assessment on the workplace with the focus on 'Operation,' whereas the main areas of ergonomics, psychology, education, and training are taken into account—Human Performance Technology (HPT)
- Directive, originally 392/89/EC, currently 42/2006/EC: Tools for minimizing risks on machines and machinery to achieve their safety and reliability.

Specific requirements to execute risk management measures are dependent on the complexity of the systems being analyzed. These systems are interconnected and controlled by central processing units within integrated management systems (IMSs) (see Chapter 3). Machines and mechanical systems not only have various properties, but are also implemented in various technologies that are currently on a high technological level.

General rules of prevention for developers, constructors, and draftsmen are stated, for instance, as the following:

a. Elimination of danger and the risk thereof
b. Assessment of risk, especially with selection and during usage of operating devices, materials, substances, and guides
c. Execution of regulations against dangers at the place of their source
d. Preference of collective protection measures contrary to single protection measures
e. Adjusting tasks according to the skill and technical level of an employee

f. Observing human abilities, features, and possibilities, especially when designing the workplace, selection of working devices, guide, or manufacturing process with the aim of eliminating or lowering the effects of harmful factors of jobs, difficult jobs, and monotonous jobs on employee health

Attention has to be paid in this regard to new methods of identifying dangers and threats with the help of new information, measuring, and diagnostic technologies. According to the risk theory (see Chapter 4), it is obvious that it is effective to identify threats at the workplace planning stage or at the stage of product design. This procedure can be executed by simulation technologies, that is, 3-D or virtual reality. These technologies allow us to prepare working activities so that it is possible to inform the user of the technology or product about the remaining threats in accordance with all of the currently enforced legislative regulations. Experts try to develop new methods in the field of risk management to quantify and minimize values of remaining or acceptable risk, related to requirements of setting this value at a certain level, which is a part of modern laws as well.

7.1 CHANGE IN MANUFACTURING TECHNOLOGIES AND CONSEQUENT REQUIREMENTS FOR EDUCATION IN OCCUPATIONAL HEALTH AND SAFETY

Change in the nature of work and working conditions due to technological innovations and globalization of labour markets, as well as demographic changes in society, presents a great challenge for occupational health and safety. This also affects the development of new forms and methods of educating employees in the spirit of the assumption that only motivated and healthy employees present a basic requirement for a company to be competitive and, at the same time, able to retain and create new jobs.

Demographic changes in economically developed countries lead to aging of the labour force and therefore to new types of risk, ones that are specific to senior employees (see Chapter 5). Companies and educational institutions, including colleges and universities, have to take into account this development and adjust their educational systems, including the syllabi of study programs, lengths of educational modules, forms of education, and so on.

Changes in working conditions affect new requirements for employees and their qualifications and even for new forms of education and the actual content of educational processes. It is conditioned by these facts:

- Labour markets have become more multinational. International corporations use the same methods of risk management. Multilingualism has gained more importance (see Chapters 1–6) and that has become a challenge for linguists.
- Utilizing IT has been constantly on the rise; employees now have to demonstrate their skills in IT, which requires thinking logically, abstractly, analytically, hypothetically, and in a planning way. Mathematical knowledge

has gained more importance, too; therefore, it is necessary to implement IT subjects into various stages of education.

- Professional and social qualifications have to continually improve; education and broadening knowledge, especially in various forms of lifelong learning programs, has been the subject of increasing investment.
- Changes in business frameworks and decentralization in the workplace require further independence, creativity, self-initiative and responsibility, knowledge and skills in the field of communication, cooperation, and teamwork. Social competencies and the ability to work in a team have become, in many cases, more important than specific knowledge.
- Jobs have become less place and time dependent. Employees are required to be more flexible and mobile.
- The population is aging. Relationships between age groups of employees change. The average age of employees has increased.

These trends of developing new technologies in the area of occupational health and safety, as well as their nature in other areas, set the requirements for quality and skills of experts in the field of safety. These questions are scrutinized in recent specialized literature, and we might state that a modern expert in the field of occupational health and safety management must be highly qualified, and able to make use of information from various fields at work. First though, one must be a generalist who utilizes a wide array of knowledge at work.

7.2 NEW PRINCIPLES OF OCCUPATIONAL HEALTH AND SAFETY

Changing forms and organization of work create new requirements for occupational health and safety, and require the application of new methods of prevention. Healthy, motivated, and less unequally tasked employees guarantee a constant quality of final products and services.

Operative application of up-to-date knowledge, such as the results of research and development, causes a dynamic progress of developing modern complex engineering systems and related manufacturing technologies. This progress also causes new types of threats (e.g. new chemical substances, use of radioactive material, etc.), which imposes increasing demands on experts' knowledge. Furthermore, it causes constant development of legislation on the national and international level. Market globalization requires reckoning with legislation of socio-political alignments and their consecutive implementation subject to conditions of specific countries of these communities, for instance, the European Union and its member states. Moreover, it raises the need for constant learning and therefore a constant feed of recent information. Recommendations of the International Labour Organization (ILO) for occupational health and safety management systems from May 2001 (see Chapters 2–4), which includes elements of some new procedures, serves as an appropriate example. It is then necessary to become familiar with its content, for instance, for any expert in the area of occupational health and safety, by attending distance learning courses. Furthermore, experts themselves should try to acquire, master, and implement this information in their jobs as soon as possible.

The aforementioned trends of developing new technologies in the area of occupational health and safety, as well as their field-extensive nature, set the requirements imposed on quality and abilities of experts in the field of safety.

Some companies include methods of prevention within the frame of occupational health and safety in their corporate philosophy and quality management. They have realized that employee satisfaction and motivation are both important business and economic factors. Safety culture is provided by all of the managers and not only one responsible employee (see Chapter 1).

The load of work in process is being changed as well due to changing conditions on labour markets and in manufacturing technologies, while new and emerging risks occur. The physical load of work is becoming less important, while the mental load is increasing. Employees frequently take over the responsibility for their own health and safety. Therefore, they must be able to handle stress and mental burdens in order to handle criticism and responsibility while still motivating themselves to work effectively. Demands on mental resistance have considerably risen in comparison to the ability to resist physical load. Mental burdens have increasingly emerged in specific groups of employees, for instance in highly qualified groups. This progress also causes new requirements to be imposed on education and training of all employees. Syllabi should include areas such as psychology, sociology, communication, and so on.

Health requirements have gained new importance. Employees want to retain their health and not be constantly threatened by their working conditions. They consider health as the essence of their performance.

Occupational health and safety as well as ergonomic aspects of the workplace must be duly taken into account in business frameworks and processes. They do not need to be improved, that is, altered, which is time-consuming and costly, if they are reckoned with as early as the design and development stage of a product or technology. Manufacturers, suppliers, and importers bear responsibility for providing the markets with safe and properly designed ergonomic machines and devices. Users are informed about the residual risks and are advised how to safely use the products. Bodies of technical supervision, as well as occupational safety inspectors, act as counselling institutions for safety of technical devices and therefore support the state in influencing occupational health and safety issues. Researching and counselling organizations offer services in the fields of occupational health and safety management and safety of technical systems as well.

7.3 EDUCATION OF EXPERTS IN OCCUPATIONAL HEALTH AND SAFETY

New methods for threat identification that can be executed with up-to-date IT methods and new measuring and diagnostic technologies must be the main point of interest in education. Herein we can use various methods of 3-D simulation, for example, in the form of virtual reality. These tools help designers and developers to execute threat analyses in the design stage, on a computer screen. These methods help to shape machine operation, as well as complex mechanical systems by executing analyses to determine the activities in which to expect threats and the risks associated with them. Scientists emphasize the development of new methods as opposed to quantification of risks, and therefore, procedures to define values of residual or acceptable risk.

A strategic educational task on the current globalized labour market aimed at the European labour market is defined in 'EU Council Presidency Conclusions,' Brussels, March 2, 2007 (04.05), 7224/1/0, point 15: "General and professional education is the basic condition for a well-functioning knowledge triangle—education, research, innovation—and contributes to the economic growth and growth of employment. The last 12 months witnessed an effective progress of implementing the program 'General and professional education 2010.' Member States are determined to continue in reforms and thoroughly implement the working program aimed above all at upgrading college education to provide superior and attractive professional education and at implementing national strategies for lifelong learning."

Even the Declaration of the XVII World Congress in Seoul in 2008 supports the importance of employee education focused on occupational health and safety: "Employers must discuss issues of occupational health and safety with their employees with the aim of creating respective conditions for their education."

What conditions the change in the system and forms of educating current experts in occupational health and safety as well as experts in safety of technical systems?

First and foremost, these are the regulations that are defined on the basis of detailed analysis of requirements for a modern work arrangement, working processes, and labour performance. Abilities to evaluate and assess used regulations amended by inspecting results related to requirements stated by developers and machine manufacturers impose high expectations for education of experts in the field of safety they follow the:

1. Progress in innovative processes, development of new technologies, and therefore new machines that cause new types of risks to emerge. To successfully manage these risks, one has to be effectively educated, attend practical courses and training for future experts, including machine designers, workplace developers, and even their users. However, experts do not possess relevant information about operational conditions, which presents a problem. Furthermore, information from manufacturers is not proven by practical experience.

2. Safety of machines and mechanical systems, as well as workplace safety, which are not nowadays systematically integrated (maybe just partially) into study programs, especially at technical colleges. Regulations to minimize risks at work are often deemed financially demanding and decrease work productivity. The goal must be a certain integrated approach where relevant subjects are implemented in syllabi of mainly engineering programs during the actual study of specific subjects or within lifelong learning programs. These subjects are aimed at areas of risk management from a broader point of view. It is effective to realize them as part of integrated management systems—quality, safety, and environment—or currently even in a complex of generic management systems. A train of thought has to be focused on reaching final product quality and productivity of manufacturing processes,

which has to be supported by respective regulations in the stage of their design and development.

3. Fact that a worldwide educational space creates conditions for executing common study programs, and the quote "safety does not know borders" is the basis for it. Therefore, it is useful to consult methods and education content with partners from educational institutions and from practices abroad with the aim of preparing an integrated educational module focused on gaining knowledge from the field of safety and on getting familiar with the basic habits to implement them in real life. It is also useful to choose a combination of scientific facts, generally formed theoretical conclusions, and practical experience with the aim to build an expert in the field of complex safety.

4. Belief that the first step in activities that aim to achieve quality and safety for machine and mechanical system production leads to creating an information database of possible network members from colleges or lifelong learning institutions that are interested in implementing practically aimed educational blocks in their study programs. Working meetings with participating representatives of industrial practice will hold discussions about various models and contents of specific educational models with the aim to draft a module that would be used within several college systems of technical fields of study.

Development in all of the fields of occupational safety places increasing demands on employees and their studies. Lifelong learning is now part of being prepared to handle changes in the labour market. It means the same for undergraduate and graduate students in colleges where safety is taught as a separate study program or as part of study plans in other fields of study. Experts in the field of occupational health and safety must provide a specialized capability necessary to solve issues of this field. However, corporate safety culture has to be assured by all of the employees who are aware of the importance and task of prevention within any company activity (see also Chapter 1).

An expert's capability must emerge from his specialized knowledge. Only then can it partner with experts in the stage of planning and design, purchase, design of suitable technological and logistic processes, selection of proper materials, production, trials, and choosing the maintenance strategy.

It is not possible in these activities for an expert in the field of safety to have the same knowledge, skills, and experience as an engineer or electrician in their respective fields, or a physicist, chemist, psychologist, sociologist, doctor, or experts from other areas, which might be part of safety analyses of machines or technical systems or even complex technologies. What's important is to have expertise in tools for risk analysis, so that the expert is able to apply certain system approaches within risk management with the aim of risk minimization. Current requirements imposed on experts in various fields of engineering include implementing new areas into engineering education. This is especially IT, mechatronics, system technology, environmental technologies, and risk management.

Current educational systems, especially in undergraduate and graduate studies and in research, are distinguished by their *openness*. This means not only the possibility of student and teacher exchanges, the exchange of scientific and specialized information, but also common solutions within international solving teams. The reason for these activities is the effort to have highly educated experts on labour markets, who are able to work in various countries where corporations have their subsidiaries as a part of the global economy. This is just another reason to gather finances for international teams to solve recent issues of research, development, and innovation.

Areas where we might expect intense international cooperation in the future will take into account the following conditions:

1. Relocation of production facilities to various countries of the world.
2. The labour market is globalized, does not know borders, and injuries are not tied to nationality.
3. Integration of laws, regulations, and directives from the field of safety into active life.
4. Creating a basis for a united safety culture.
5. Respecting some marginal conditions as language and written text in a company.
6. Providing a single communication regardless of different cultures.
7. Internationalization of research.

Education of experts in occupational health and safety and in safety of technical systems can be done in the future, either by college education or by lifelong learning, for instance, through these two models:

1. In a specialized study program with a working title of "safety of technical systems and occupational health and safety" in undergraduate, graduate, and postgraduate level (Figure 7.1)
2. Through postgraduate studies or within a lifelong learning program after graduating from a mainly technical- (or natural science)-oriented study program (Figure 7.2)

It is expected in a modern society that the safety of any product or technology is a high-priority goal. It is important for these study programs to use knowledge and experience from classic engineering fields of study such as mechanical engineering, production technologies, construction, extractive metallurgy, and electrical engineering.

New risks emerge because of the globalization of labour markets and development of new technologies. Requirements imposed on emerging international labour markets are conditioned by the integration of European legislation into the legislation of the respective member states. This creates conditions for cooperation in the field of research and college education. The European Union announces project appeals in order to make this process more effective. These projects, funded by the EU, then facilitate the creation of networks between relevant institutions in the fields of research and education. First and foremost, the projects are focused on creating conditions for a single 'European' safety culture in corporations.

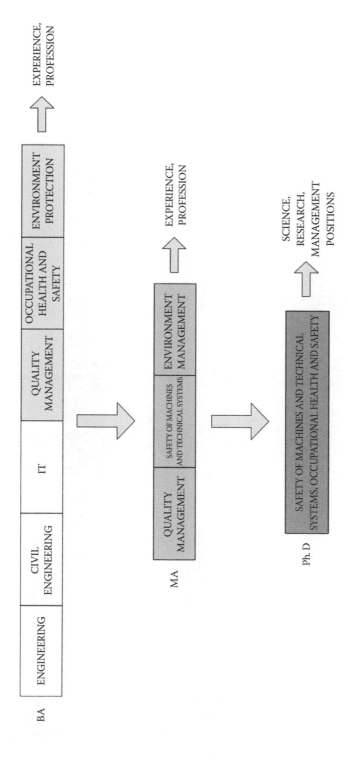

FIGURE 7.1 Model of specialized education.

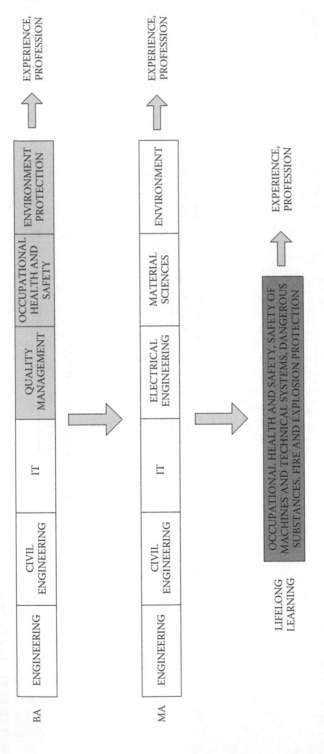

FIGURE 7.2 Model of postgraduate studies.

The amount and content of newly created requirements imposed on the final product and on the production process have also recently become very important and extensive. In 1992, the need to establish an independent education and scientific research unit of the Department of Safety and Quality of the Faculty of Mechanical Engineering at Technical University of Košice was recognized. The unit's main task is to provide a learning process, within basic graduate studies and within specialties, and to take part in various forms of lifelong learning. Currently, the main activity is to produce college-educated engineers for the field of safety of technical systems, occupational safety, and production quality directly related to environmental science. The Department of Environmental Science and Industry Management guarantees the education process and research for the field of environmental sciences.

7.4 PARTIAL CONCLUSION

New approaches in the system of occupational health and safety management require everyone to be aware of the risks that one must live with, in the workplace and in everyday life as well.

Every employer is responsible for identifying risks in the working process, carrying out measures for their elimination or minimization, and informing employees about any possible residual risks.

In order to fulfill these responsibilities, it is necessary for managers and specialists in the field of occupational health and safety management to know the trends in the field and to lead by example in their profession. Their task is to influence the implementation of and adherence to the rules of occupational health and safety, with the help of information, communication, training, and thorough supervision. It is expected that they will associate themselves with the principles of occupational health and safety management and that they will be willing to listen to their coworkers' opinions. To execute these activities, it is absolutely necessary that they systematically study and provide education for all of the workers they oversee.

Globalization of labour markets, development of new machines and technologies and the emergence of new risks, changes in demands on working conditions, and interconnected legislative regulations of various countries create opportunities for cooperation, with risk management as part of educational and research projects regardless of state borders. Various types of projects to facilitate common education programs and research are created with international cooperation to speed up these processes. Common education programs help to create conditions where everyone is familiar with the same knowledge of risk management, regardless of diversity of countries. At the same time, they contribute to bringing specific national safety cultures closer to a common safety culture within the European Communities in the spirit of the following statement:

> When implementing modern methods of management, it is not possible to use procedures of yesterday and knowledge of the day before.

BIBLIOGRAPHY

Sinay, J., 'Projekt post diplomového vzdelávania inšpektorov a expertov pre bezpečnosť a ochranu zdravia pri práci v rámci programu Európskej únie', Conference: Current Issues of Occupational Safety, Occupational Safety Institute of Research and Education, Bratislava, November 1999.

Sinay, J., 'Die Rolle von Wissenschaft und Lehre beim Risikomanagement', Konferenz 2000, Zukunft—Arbeit—Preväntion, IVSS Sektion Maschinen- und Systemsicherheit, SUVA Luzern, Switzerland, May 2000, pp. 224–242.

Sinay, J., 'Znalosti BOZP: Neodmysliteľná súčasť manažérskych zručnosti', XX Conference of Current OHS Issues, Starý Smokovec, 2007.

Sinay, J., 'Sicherheitsforschung und Sicherheitskulturen', Transnationales Netzwerk-Symposium, Bergische Universität Wuppertal, NSR, 29–30 October 2008.

Sinay, J., 'Anforderungen an eine moderne', Arbeitsschutztag Sachsen-Anhalt 2010, Landesarbeitskreis für Arbeitsicherheit und Gesundheitsschutz in Sachsen Anhalt. Otto von Guericke Universität Magdeburg/SRN, 2010.

Sinay, J., and Bartlová, I., 'Zmeny výrobných technológií a z nich vyplývajúce požiadavky na vzdelávanie pre bezpečnosť a ochranu zdravia pri práci', in *Bezpečnost a ochrana zdraví při práci*, 2011, sborník přednášek, 11, ročník mezinárodní konference, Ostrava, 10. květen 2011, Ostrava: VŠB - TU, 2011, pp. 177–183, ISBN 978-80-248-2424-6.

Index

Printed and bound by CPI Group (UK) Ltd, Croydon, CR0 4YY

18/10/2024

01776264-0001